Praise for Impactful Inclusion Toolkit

"Yvette Steele has created one of the most practical, useful, and comprehensive guides to enable everyone—regardless of title, department, or status—to be powerful change agents for inclusion in their own organizations. Steele has gathered a wealth of resources, tools, and real-world examples to illustrate to readers what true inclusion looks like in action. Steele then takes this further by providing actionable, hands-on steps for readers, which helps to solidify and deepen the reader's understanding of their learnings. This is an invaluable resource for leaders seeking to create safer and more inclusive work environments for all."

Susanne Tedrick, author of
Women of Color in Tech

"*Impactful Inclusion Toolkit* by Yvette Steele is a powerful tool for starting some difficult and necessary conversations. Happily, there are 52 activities, so you can tackle one a week and not try to address all of these topics at once.

"This book is emerging at the perfect time. Events in politics, race relations, and the global society have converged in such a way that the world is ready to begin the hard work of addressing diversity and inclusion at a higher level than ever before. Yes, we've talked about these issues before but often from a watered-down perspective. Some people got a pass. Compromises have always been made. Yvette Steele's work breaks through all that and proclaims that the time is now to stop giving passes and making compromises and to start addressing the reality of exclusion that permeates our institutions and society. *Impactful Inclusion Toolkit* is not a vague call for awareness or action. It is a thoughtful and powerful tool for starting the conversations that need to take place across the country and around the world. The last few years have been very difficult when it comes to addressing matters of diversity, equity, and inclusion. But those tough times made it possible for a work this powerful to emerge and to help people get past the rhetoric and politics to work on the actual details of making the workplace a welcoming place for everyone."

Karl W. Palachuk, author, coach, and community builder

"I highly recommend every business leader pick up this book and use it as a guide. Yvette Steele has structured this resource to be highly actionable with ways to create an inclusive work environment and shift your workplace culture. It is not a one-and-done training guide but provides practical steps through prescribed activities that will help you self-assess as well as promote inclusion over time. I recommend you give one to every employee as part of their new-hire toolkit. It helps both individuals and leaders to find their own way and adjust their own habits and practices to behave with intentional inclusion. As a leader, I have read and participated in many trainings and guides. None was as practical and insightful as this one."

Gavriella Schuster, former Microsoft corporate vice president and DEI advocate

"*Impactful Inclusion Toolkit* lives up to its promises of helping anyone at any level create a more inclusive workplace. Whether you are just starting your inclusion journey or have been on this road for a while, Yvette Steele lays out a roadmap of activities, follow-up actions, and resources to help us understand what inclusion looks like for people of color, members of the LGBTQ+ community, individuals with disabilities, and others, as well as provide practical steps we can take each day. I highly recommend this resource for all who want to create a more inclusive work environment and society."

Kathryn Rose, founder and CEO of getWise

Impactful Inclusion Toolkit

Impactful Inclusion Toolkit

**52 Activities to Help You
Learn and Practice Inclusion
Every Day in the Workplace**

Yvette Steele, Founder,
DEI Insider

WILEY

Published by John Wiley & Sons, Inc., Hoboken, New Jersey.
Published simultaneously in Canada and the United Kingdom.

ISBN: 978-1-119-93020-4
ISBN: 978-1-119-93022-8 (ebk.)
ISBN: 978-1-119-93021-1 (ebk.)

For general information on our other products and services or for technical support, please contact our Customer Care Department within the United States at (800) 762-2974, outside the United States at (317) 572-3993 or fax (317) 572-4002.

If you believe you've found a mistake in this book, please bring it to our attention by emailing our reader support team at wileysupport@wiley.com with the subject line "Possible Book Errata Submission."

Wiley also publishes its books in a variety of electronic formats. Some content that appears in print may not be available in electronic formats. For more information about Wiley products, visit our web site at www.wiley.com.

Library of Congress Control Number: 2022940152

Cover image: © master1305/Adobe Stock
Cover design: Wiley

This book is dedicated to all those who choose to blaze a trail to equity for all despite the cost. Together we can change the world one word, one behavior, one practice at a time. I am on this journey with you. I've made mistakes just like you have or will. I celebrate your wins and applaud you when you get up from setbacks. Stay the course. See the impact. The world needs role models like us. This is how we get change done.

About the Author

Yvette Steele is an equity accelerator and inclusion strategist who believes in the power of living authentically at all times. She draws on her decades of experience to help organizations realize the promise of a diverse work-force by implementing strategies that eradicate the barriers many face in maximizing their potential and empowering people to integrate inclusion into day-to-day interactions. She was recognized as a diversity thought leader on the inaugural Channel Futures DE&I 101 List for advancing diversity, equity, and inclusion through words, actions, and leadership, and she has served on the DEI Advisory Committee for YMCA-USA, Tech Advisory Board Steering Committee for the National Urban League, and the DEI Task Force for Chicagoland Chamber of Commerce. You'll find Yvette sharing her leadership and vision on many diversity- and workforce-focused podcasts, blogs, news articles, and panel discussions, including the *Wall Street Journal*, the Society for Human Resources Management (SHRM), and the ChannelPro Network. As founder of DEI Insider, she develops resources that help people practice inclusion every day and supports members of historically excluded groups to overcome adversity to thrive in the workplace by providing tips, tricks, and best practices. A native Chicagoan and graduate of Chicago State University, she lives in Chicago with her husband and children.

Acknowledgments

Thank you to my awesome husband who has always believed in and encouraged me through ups, downs, and in-betweens. He shows me real love each day and never takes a day off.

***

Thank you to my children who gave me the courage to awaken each day in the darkest of times. You have and always will be my inspiration.

***

Thank you to my brother and sister-in-love who have surrounded me with unconditional love.

***

. . . and finally to my mother, father, and sister who motivated me in unimaginable ways.

Contents at a Glance

Contents

Foreword

The explosive growth in interest of the diversity, equity, and inclusion field over the years is unsurprising to many, particularly as we all have watched a number of events across the globe that highlight the continued oppression of marginalized groups. From the waves of anti-Black and anti-Asian violence to the promulgation of laws authored with the intent of excluding the LGBTQ+ community, examples of inequitable and exclusionary practices are many. It is this collective societal reckoning with diversity and our complicated history with various identities that has driven this wave of investment in new DEI programs and business functions.

Yet even with an abundance of new DEI certification programs and the like, we as practitioners still find ourselves faced with a familiar set of questions from those who are looking to apply these principles to their daily lives: What do I do? How do I make a difference? How do I demonstrate my commitment to DEI and to people who are different from me?

It's a persistent set of questions that practitioners like me in the corporate space face regularly. While these questions generally come from a place of genuine curiosity, I also acknowledge that they are often driven by fear. Many people do not want to mistreat those around them and certainly fear the perception of being any type of "ist" in a world where what has been labeled as "cancel culture" (also known as accountability) has found prominence. Still, these questions persist not because the tools we provide are impractical or even extremely difficult to apply.

Rather, it's usually because there is additional foundational, introspective work that the learner must do, and worse yet, there is little to no instant gratification. Transformational change at both the individual and corporate levels takes time and consistent investment. The flawed view that one can just call themselves an ally because they have that one gay friend or they stopped a woman from being interrupted in a meeting that one time is just not going to cut it. Learners and organizations alike must actually unlearn behaviors and practice new and inclusive behaviors to the point where they become habitual and normal.

Learning in the DEI space is plagued with cursory modules that focus only on anti-harassment and compliance, especially as it relates to the workplace. Don't misunderstand me here; the content of those modules is important and should be used to continue to address some of the ills of toxic and noninclusive corporate cultures. But that alone does not speak to the everyday needs of a learner or give them the opportunity to think about the ways that identity impacts how they navigate the world. More important, most corporate diversity training does not provide learners with the opportunity for ongoing practical application.

It is in this gap where the author finds her audience. Yvette Steele grounds her work in both authenticity and practicality for the reader. Her authenticity—and of course her ability to sell me an idea—is what sparked our connection years ago. She has been a great collaborator and partner as I have taken on more expansive roles building both the DEI and environmental, social, and governance business functions in global Silicon Valley tech organizations. I have been an excited supporter and observer—and hopefully somewhat helpful—as she has really turned her focus to diversity, equity, and inclusion work. Leveraging her own personal journey, multiplicity of identities, and investment in learning inclusive and equitable practices, she takes the reader on a journey that begins with self-awareness and leads them to behavioral changes.

As you move through the activities herein, you are not so subtly encouraged to pause, reflect, and even role-play—to really think about what you would do in any of these scenarios. And that is coupled with small yet impactful actions that individuals can take in the course of their daily lives to effect change. She encourages you to take this in small bites, focusing on a couple of activities at a time and doing the necessary reflective work to truly understand what you read and what it looks like in practice. If you are a seasoned practitioner in this space, this book is rightfully not written for you. This is for those who have found themselves looking for ways to get started on a personal journey of re-learning.

It is important to note for any work focused on DEI practices that things change over time. As people become more comfortable identifying themselves in various ways both in and out of the workplace, and societal norms change, so too will these scenarios and activities. That makes this a great, living piece that can be revisited and offers readers the chance to also examine the field's growth over time alongside your personal development. Stay the course, dear reader. The journey is a long and difficult one, and you are given a great guide to help you along the way.

Ulysses J. Smith
Global DEI strategist and executive

Introduction

Have you ever found yourself wondering "What can I do to support diversity, equity, and inclusion (DEI) in the workplace?" I believe that at some time or other, many people ask themselves this question. Just imagine a place that you helped to create where everyone feels included and empowered to thrive in their authenticity. That's the place where everyone wants to be, and that's what we're striving for. Inclusion starts with the letter *I*—meaning that it starts with you. You begin by being Intentional. There's that *I* again. Your consistent intentional actions lead to impact or as I like to call it "Impactful Inclusion."

This book is born from my desire to help my many friends and colleagues who have approached me for guidance on how they can be a part of the solution to the inequities taking place not only in the workplace but society as a whole; and I love them for asking. Thank you for sparking this idea and creating this opportunity to share my personal experiences as well as those of family, friends, and colleagues.

There are hundreds of DEI books written for leaders and managers—people with positional authority. However, few books focus solely on the everyday person—the individual contributors who keep businesses humming. No matter their job role, everyone either experiences or perpetuates exclusion in daily interactions. We are often left without the tools or knowledge needed to recognize and address exclusive acts and replace them with acts of inclusion. My goal is to get this book in the hands of 10 million people for the good of all humanity. I believe that when we know better, we can do better. We can build a world where historically excluded people are included and valued.

Regardless of your position within an organization, you can incorporate the activities found here into your day-to-day interactions for better relationships, greater gratification, and profound business outcomes. You can be a catalyst for change. The key to success on your inclusion journey is to pack an open mind, receptivity to change, and willingness to learn. While these activities are simple, they may not always be easy. Inclusive behaviors are a skill, and like any skill, they will improve over time with practice. Lean into the discomfort. You're going to make mistakes. It's okay. It's how you learn. Stay with it. Visit my website, DEIinsider .com, for continued insights and resources. Join others on the journey by engaging in our LinkedIn group "Champions of Inclusion."

The Merriam-Webster dictionary has several definitions of inclusion. I was thrilled to see this one: "the act or practice of including and accommodating people who have historically been excluded (as because of their race, gender, sexuality, or ability)."

I remember a time when inclusion simply meant the act of being included. We've evolved into a society with an increased awareness and sense of responsibility to make the world inclusive for all. Throughout the book, I use the term "we," referring to you the reader, and me the author, as I am working to become more inclusive along with you. Thank you for joining me on this journey.

How to Use This Book

Let me begin by saying that this is not a quick read over the weekend; rather, it should be taken in weekly doses where you make time to acquire a new skill to practice every week until it becomes a habit. Journal about your reflections, what you've discovered about yourself and others. Capture successes and lessons learned. Celebrate wins.

Inclusion is everybody's job that anybody can do. Place this book somewhere where it will stay top of mind. Set weekly reminders to build your inclusion muscle. Each week, work to incorporate the actions into your day-to-day interactions and review the action accelerators to increase your understanding. My hope is that your successes and energy will lead to others joining you. Journeys are always more fun when you share them with others. There are 52 activities. Start with the first one of getting to know yourself. When you get to activity number 52, begin again at 1, and note how much you've evolved. Imagine the impact you'll make year after year.

Commit to actively changing thought processes and behaviors that ultimately create more inclusive work environments. Above all, enjoy the journey!

My Battle with Exclusion

I grew up during a time where societal norms judged and frowned upon divorcées, single motherhood, working mothers, men marrying women with children, and other ridiculous beliefs that in effect stigmatized women like my mother. Society got to decide who was acceptable and unacceptable or worthy of respect. My mother was raised in poverty in the slums of Chicago by a single mother and never met her father. A blend of German, African American, and Cherokee Indian, her green eyes, French vanilla skin, and long sandy brown hair made her stick out like a cherry in a bowl of milk. I recall the dozens of stories she shared, lamenting about the bullying at school, or the summers spent in Minot, North Dakota, with White family members, and the impact of constant reminders of the darkness of her skin. By the age of 25, she had been through two divorces and had two children. I was three years old and the youngest when she and my father went their separate ways. But I have to give it to her: Mom was a strong and determined woman. With the support of her mother and uncle, she attended nursing school while my grandmother cared for me and my brother. Looking back, I can only imagine the scars and the emotional trauma she endured and carried her entire life just trying to fit in. Those experiences definitely informed the way she cared for her children. She was determined to be accepted by society no matter the cost.

My first brush with exclusion happened when I was five. My mom took me and Tonia (not her real name), the five-year-old daughter of her best friend out for shopping and dinner. She dressed me in an outfit that made me feel like an old lady. I'll never forget that hideous blue ruffled front dress with lace-top ankle socks that kept sliding down into the back of my black patent leather Mary Jane shoes. I wore thick cat eyeglasses, and my hair was pulled into a wild, bushy ponytail. I envied Tonia. Her hair was pressed bone straight and cascaded just past her shoulders. She wore the cutest jumper and go-go boots. And my mother? Well, stunning. Her hair swept into a French roll, and she wore a knee-length sheath that not only accentuated her curves, but it was also a shade of green matching her eyes. Everywhere we went, compliments poured about the beauty of her daughter Tonia and how much she looked like her mother. Mom beamed at every flattery while I waited for her to correct them and identify me as her daughter. By the third instance, my eyes welled with tears. I could not understand why I wasn't her daughter that day. I did not realize until I was well into my 40s that it was her desire to fit in with the picture-perfect child that drove her decisions to not correct the mistaken identity.

She remarried when I was seven. Soon after, my brother started living with his paternal grandmother. I was told that he chose to leave us because his grandmother promised him a dog. Mother explained that she wanted no one who desired a dog over her, so she opened the door and allowed him to leave at only 11 years of age. I felt abandoned. How could he leave me as well for a dog? Why couldn't I go too? I don't recall him ever saying goodbye. He just left for a weekend visit with his grandmother and never returned. I now believe that the stigma of a ready-made family led to my brother's quick departure. Alone, I tried hard to be the perfect daughter for my mom and stepdad. Turns out, I was awful at being perfect and was criticized or punished for every spoken and unspoken rule I ever broke. I maintained a relationship over the phone with my brother for a few years as best I could, but eventually that wasn't enough, and we grew apart. Three years into the marriage, my sister was born. Mother blamed me for having to give birth to another child because I could not bring myself to call my stepfather "daddy," something he so desperately wanted. For me, it felt too awkward to say "daddy" to someone with whom I had no emotional connection. I knew my father, and *he* was daddy to me. My sister was my stepfather's first child and would ultimately become his only child. The excitement from his family members around her birth brought frequent visitors and loads of gifts. Often, these visits relegated me to another room rendering me invisible. Seemingly, no one wanted to engage with the child from the previous marriage. I remember stepping out once to see all the presents, and when I attempted to touch one, I was scolded by one of his relatives. There was one relative who visited for years every weekend bearing gifts, none of which were ever for me, and she made that clear. By the time I was 11 years old, I was made to feel unlovable regardless of how hard I tried. I could not remember the last time I saw or even spoke to my biological father. Attempts to connect with him were forbidden because he never paid child support. The photos of him that I had hidden in a drawer were discovered and destroyed. It's as if my mother wanted to wipe out his existence. Except, there I was, a constant reminder of her past.

In the midst of struggles to fit in at home, I still had a privileged life. I attended private schools off and on, never missed a meal, took family vacations, and always had plenty of Christmas and birthday presents. We were a picture-perfect family.

I grew up in an area that was transitioning from White to Black residents. My first encounter with racism happened in the fourth grade. I walked to a Woolworth store with a friend who was also Black. Back then they had over-the-counter food service, and we wanted pizza. The White woman behind the counter took what seemed like forever to acknowledge our presence. So my friend finally blurted out that we wanted to order two slices of pizza. The first thing the woman said was, "Do you have money?" Proudly we placed our crumbled bills and loose change on the counter. The woman counted it down to the last penny, looked at our excited faces, and said in a condescending tone, "You don't have enough." We knew how to count, and we knew that we had enough money; but we had been raised not to challenge our elders, so we did not question her. Determined to have our pizza, we then requested to order just one slice to share. She told us again that we didn't have enough money. The excitement melted from our faces and was replaced with sadness and confusion. We scooped our money from the counter and left. For weeks I was troubled, trying to figure out what happened and wanted to ask for help, but I didn't know how to articulate it. A part of me was afraid that I would discover that somehow it was my fault and that I would be punished.

In public grade schools, I was bullied for being either taller than most, light skinned with long hair, or speaking proper English. Once again, it was hard for me to build relationships due to things that I could not control. Speaking slang was reprehensible in mom's house, so I seldom used it unless I was certain that she would never find out. The so-called friends I did make played with me only to "borrow" my toys and clothes that they never returned. In the private grade school I attended, I was subjected to mean girl pranks long before it became a popular movie. In the eighth grade, just after spring break, someone started a rumor that I had sexual intercourse with a boy in class who was incidentally ostracized because he had the darkest shade of brown skin. It was what most Black people refer to as blue-black. The rumor claimed that I now had cooties and warned everybody to stay away from me so that they would not catch it. Everybody in the class heeded the warning. I had no idea why I was being avoided at every turn—in the lunchroom, on the playground—and was uninvited to parties. A classmate secretly told me what was going on. She shared how she didn't like the prank and pitied me yet blamed me for it. This treatment lasted to the end of the school year.

By high school, I still had no best friend to call my own. I was the one who was busy being a friend and expecting friendship in return that never materialized. My first year of high school was totally wild. I had never been around hundreds of teens from so many environments. It was a predominantly African American public school accessible to kids from low as well as the upper middle-class backgrounds. I struggled to find my tribe. Cliques were already well established as the friendships made in grade school continued into high school. While striving to fit in somewhere, I met my now lifelong friend Jacqui. We struggled together, but she was better at fitting in than I was. When we hung out, bullies bypassed her to pick on me. I never understood why. I just figured that she was a cool kid and for some reason or another must have thought that I was cool too. One afternoon I cut class and ran into a group of classmates smoking marijuana. When they asked me to join them, I jumped at the chance. This was my opportunity to make friends. I had heard the term "Cloud 9" before but had never been there until that day. Cloud 9 was a place of extreme bliss. I had no pain, no cares, and felt totally free. I could escape it all for a few hours, and I never wanted to leave. It wasn't long before mother figured out my new bad habit and newly found friends and assured me that the following year I would be back in private school, and oh, by the way, she wasn't going to pay for it. I had better get a job or find somewhere else to live. At 14 years old, she advised me to claim to be 16 years of age on job applications, as that was the minimum age required for employment. It worked. I landed my first job in a few short weeks so that I could pay my high school tuition. Back in the 70s, all one had to do was write their age on the job application, and no one verified it like they do today.

Sophomore year, I entered a private school on the near north side of Chicago and worked afternoons and weekends at a fast-food chain just blocks away. This school was very diverse. Not only did students converge from various socio-economic backgrounds, but they were also other cultures and races—and everybody loved to party. I had finally found my tribe, and most of them were of another race! I was never bullied and was considered one of the cool kids. Working after school and weekends kept me out of the house for sanity's sake, with the financial freedom to support my bad habits. Home life was still miserable. After a full day of school, a four-hour work shift, a ninety-minute commute on public transportation, I returned home to a set of chores.

Keeping the entire house clean was another full-time job that I deeply resented, but it was how I earned my keep of food on the table and a roof over my head. The fact that I was exhausted by the time I got home and that I had homework to do was not her concern. I needed to figure it out. I was disillusioned into thinking that because she wasn't covering any of the costs of my education and no longer supplying an allowance that the cost savings would extend me some grace. My friends nick-named me Hazel, the namesake of a 1961 sitcom about a live-in maid. When the time came for us girls to hang out, I was not permitted to leave until every chore was completed to mother's satisfaction, so they often arrived early to help me clean. I'm sure that they didn't know it at the time, but for the first time, I felt like I belonged. What mattered to me, mattered to them, and come hell or high water, we were leaving together for a night of fun.

As high school graduation grew closer, my mother insisted that I start college immediately and not wait until the fall semester. She made the choice easy. Either college or move out. And, oh, by the way, tuition was again, my responsibility. She asserted that no one paid for her edu-cation and literally said to me, "Yo ass ain't no better than mine." By now, my sister was nine years old, and I could not help but notice the stark contrast in her life as a nine-year-old and mine at that age. By all accounts, she was living her best life. She had no chores, a father who was present, and she got the adoration and accolades from our mother, while I was continuously criticized, especially after the discovery of my love for marijuana. Apparently, while doing all that was expected of me—cleaning house, working, and going to school while maintaining a B average, I still could not get anything right. At this point in my life, my brother had been erased. Our mother was now claiming to have had only two daughters when she talked about family with neighbors, acquaintances, and co-workers. The drug of acceptance is powerful. In my mind, she wanted no ties to her past. The fear of someone connecting the dots of three children, three fathers, and three marriages made space for judgment and possible rejection. Today, many people consider those situations as just a fact of life. Things happen. We now live in a time where the attitude about these circumstances is "Yeah and so what?" I wish that mom could have been that strong and not feel forced to live her life on society's terms. I wish someone had been there for her to help navigate the world she lived in.

College life was not much different than that of high school. I still lived at home, worked a full-time job during the day, and attended classes at night. I continued to be Hazel, and my friends had now moved on to their next chapter of life, either away at university or in the workforce full time. One of my jobs during college was a marketing assistant at an ethnic hair care company in the early 1980s. I was ecstatic to be a part of a prominent company in the industry founded by African Americans. By the end of my second year, I was fried. I found it to be a cutthroat environment plagued with individuals with a dog-eat-dog or crabs-in-the-bucket mentality. I experienced and witnessed exclusion based on classism (preference toward those of a certain class), colorism (a prejudice based on skin tone), and chauvinism (a male attitude of superiority over women). No one played by the rules, and the rules didn't apply to all. I've always prided myself on my sense of style and fashion. In any other environment, I was complimented for it. Here, I was labeled bougie and asked how I could afford my style of dress on my salary. People insinuated that I must have had a sugar daddy and treated me as such. I was ostracized and envied based on stereotypical beliefs. My manager, a woman about my age, never missed an opportunity to point out my mistakes and embarrass me in meetings. When I confided in a co-worker who I thought I could trust, I soon learned that she betrayed my confidence. Who knew that a toxic work environment could be created by Black people and perpetuated by Black people on one another? I left vowing to never work for "my" people again. (That said, my posture has since evolved to view that as an unfortunate situation and not a fit for me.)

After graduation, with a bachelor's degree in business administration I pursued an advertising career. It was my dream. I applied to every local advertising agency listed in the phone book and monitored the newspaper for job openings. For the first two years after graduation, I checked every few months and kept applying. I never got called for an interview and eventually gave up. I have since learned that I was fighting a losing battle as the advertising industry has a longstanding lack of racial diversity. While I don't have an ethnic-sounding name to signal my race, the college I attended and my address were clear indicators that I was not White. Eventually, I landed a sales job in a Fortune 500 tech organization. I was the first in my family to work in corporate America. My parents worked for the City of Chicago. My mother's advice for success was to have only one martini during the two-martini lunches and that bullsh!t runs the world—be prepared to give it and take it. I wasn't sure what she meant or how to apply it. Mom was not the type to explain herself. I just figured the meaning would reveal itself in time.

Six months in, I discovered that the grass was not greener on the other side. In fact, the green grass was loaded with minefields. I was the first and only Black woman on my team—maybe even the company. For the first few years, there was no one who looked like me, anywhere. I was a fish out of water with no lifeline. I had no experience in this game of sales or office politics. I was completely oblivious to the fact that I was never going to be accepted, valued, or even fit in. I was the brunt of jokes, ridiculed for carrying a purse, made to second guess every decision, and found myself constantly fighting off sexual advances. Mom didn't raise a wuss. I hung in there for seven years, and those scars compounded the ones I already had.

The school of hard knocks followed me to each new opportunity for decades. The one thing that I loved about sales roles was that as long as you made quota, you were allowed to stick around. However, I never understood why I could never be promoted to management despite my stellar numbers. Those opportunities went to White men and women who were often less qualified. I was once passed over for a management role for a White guy on a performance improvement plan. I think that's when it finally hit me—the system was rigged. No one ever intended for me to succeed or even cared about my goals and aspirations. I felt helpless but refused to give up. I am still here, fighting against exclusion.

I have come to realize how my experiences shape my view of the world and my participation in it. I am painfully aware of how being different interrupts the status quo and the lengths that people will go in efforts to maintain it. I choose to no longer be manipulated into thinking that somehow I am the cause of my own exclusion—that I am not good enough, smart enough, or something else enough. I now understand that I live in a world that was designed to exclude people like me from achieving their full potential. I accept that I have earned every opportunity that I choose to pursue while being keenly aware that I will face obstacles that others bypass by virtue of who they are.

Many may believe that there's no problem with exclusion. Everybody does it either intentionally or unintentionally. Parents do it to children, teachers to students, managers to direct reports, service reps to customers, the list goes on. Everyone has experienced exclusion at one time or another but at varying degrees. All can agree that it's never a pleasant experience where our livelihoods and happiness are concerned. Exclusion advantages some to the detriment of others. Although this is a book about inclusion, we can't address it without first understanding the impact of exclusion. It is acts of exclusion in the form of racism, prejudice, and discrimination that ushered us to this critical point in history as the cries of marginalized groups have reached fever pitch and can no longer be drowned out. The fact that we have the power to exclude through our actions means that we also have the power to include. It is through awareness and education that we can raise our consciousness on who we exclude, how we exclude them, and how to pivot toward inclusion. Deliberate decisions made throughout history disproportionately reward existing privilege and have built systems of unequal and unfair treatment. It is the outcome of these decisions that demand our intense efforts to be inclusive. Our intentional, everyday individual and collective acts of inclusion can dismantle the society that exclusion built, which continues to support those with power and unearned privilege. This book is a call to action to commit to no longer perpetuate exclusive systems and to live out that commitment through ongoing education, openness to critique, and adapting behaviors such that all can thrive. I hope to awaken on my 100th birthday in 2060 to a world that we've improved together where there is less struggle for equity and more love for the human condition.

Know Thyself

"To know thyself is the beginning of wisdom."

—Socrates

***

You may have heard the phrase "Diversity is a fact; inclusion is a choice." Think about that. Diversity is a fact of life. We are all different. As soon as you have two people in the room, you have human diversity. Inclusion, on the other hand, can happen only when the people in the room intentionally choose to value and respect one another's differences. Diversity in the workplace is about intentionally hiring individuals from underrepresented groups with the right skills and competencies in support of organizational goals. Inclusion is about empowering that blend to work together toward a shared goal, where difference is considered a strength and cultivated toward the good of the individual as well as the organization. Each day we are faced with decisions on how best to collaborate, deal with conflict, and show appreciation for one another. It's a constant process of choosing consciously or unconsciously words and actions where others are made to feel either included or excluded. Engaging across differences creates both challenges and opportunities.

We must learn to meet people where they are, including those who don't appreciate working with people unlike themselves as we work toward creating a sense of belonging and building an atmosphere of community. We start to make mindful choices during interactions and sometimes in challenging situations where we must be wholly present and reflective, while calibrating and recalibrating between comfort and discomfort. This requires building competencies to connect while creating a safe space and knowing how to pivot in critical moments of conflict to make better choices in the next—choices that enhance the relationship and achieve the desired outcome.

Communications are improved when we have the agility to go from being in the conversation to looking at the conversation. Imagine that you are earshot away from a conversation between Tamika, who is African American, and Gabe, who is European American. They are working on a high-level project and are struggling to meet the deadline. Neither is taking responsibility for the missteps that have brought them here. Gabe has always second-guessed Tamika's work and believes her to be pushy. Tamika views Gabe as an Ivy League know-it-all who believes that good ideas originate only from men. When Gabe introduces an idea, Tamika criticizes it. When Tamika begins an idea, Gabe interrupts. This back-and-forth goes on for about 15 minutes. If they were thinking inclusively and interested in solving the problem based on their unique perspectives, one or the other would eventually stop this interaction, analyze the process, and determine how to improve it. Gabe may say something like "I appreciate how you evaluate and quickly see the downside of ideas. Your feedback is good in helping us to avoid errors. Going forward, can we look at the positives as well? Tamika might say, "Gabe, it's really distracting when you won't let me complete a thought. I feel like you don't value what I have to say and find it difficult to complete my train of thought without becoming defensive. Please let me complete my statements before you respond." When Tamika and Gabe step back and look at the process and the patterns of the conversation, which are often based on beliefs, experiences, and the lens through which we view one another, they can begin to change the dynamic of the dialogue. Their stereotypical beliefs and generalizations were lurking in the background, which not only created a contentious exchange but also negatively affected their productivity.

To be effective on this terrain, you've got to know yourself, your inner self. What are your beliefs? What do you value? What triggers cause you to instinctively react in a certain way and why? In a post from Betterup.com, psychologists Shelly Duval and Robert Wicklund define self-awareness as "the ability to focus on yourself and how your actions, thoughts, or emotions do or don't align with your internal standards. If you're highly self-aware, you can objectively evaluate yourself, manage your emotions, align your behavior with your values, and understand correctly how others perceive you." The trip down self-discovery lane is well worth the time and energy and rewards you with the following:

- Power to influence outcomes
- Better decision-making, leading to a boost in self-confidence, which leads to communicating with clarity and intention
- Ability to understand things from multiple perspectives
- Freedom from our assumptions and biases
- Ability to build better relationships
- Greater ability to regulate our emotions
- Decreased stress
- Increased happiness

As we aspire to be champions of inclusion, we must put in the work to develop more self-awareness and better understand our own perspective. When we are at peace with what influences our choices around being more inclusive and develop the capacity to move back and forth from our point of view to appreciating that of others, our mindfulness evolves. Then we can begin to accept that our tried-and-true assumptions are just one of many possibilities.

Actions

Create Mindful Interactions

We are a culmination of our experiences. Those experiences inform how we build relationships. During interactions:

- Focus your complete attention on the present moment.
- Be mindful not to allow preconceived notions and judgments drive the conversation—see the other person as they are, not who you think they are.
- Be open to new possibilities, ask questions, avoid assumptions.
- During stress or conflict, remember to breathe, step outside your feelings, focus on the goal, and assess how to alter the dynamic toward mutual benefit.

Consider the Dynamics of Interactions

Reflect on a time where you met someone for the first time who is different from you. You each bring a story created by personal experiences, as well as culture and upbringing that shapes your beliefs and values and yet may or may not facilitate a connection. Consider the dynamic of the interaction where the connection was *not* made. Acknowledge the many aspects of yourself that you brought to that initial encounter and the ways the other person was different from you. Explore how beliefs, values, and assumptions may have interfered. Reflect on whether being open to and discovering the other's perspective may have changed the outcome. Note what you've learned. Contrast this experience to a time where your initial encounter with someone different from you led to a connection.

Enhance Your Social and Emotional Intelligence Skills

Emotionally and socially intelligent individuals are good at understanding how others may feel but are also adept at understanding their own feelings. Create a library of online courses, books, podcasts, etc., for continuous learning and practice. Here is a list to get you started:

- **Podcast**: *Living and Leading with Emotional Intelligence: Conscious Leadership*, with Matt McLaughlin: podcasts.apple.com/us/ podcast/living-and-leading-with-emotional-intelligence/ id1516136305?i=1000546251943

- YouTube.com: *6 Steps to Improve Your Emotional Intelligence*, Ramona Hacker, TEDxTUM: www.youtube.com/watch?v=D6_J7FfgWVc

- YouTube.com: *The Power of Emotional Intelligence, Travis Bradberry,* TEDxUCIrvine: www.youtube.com/watch?v=auXNnTmhHsk

- Udemy.com: "Emotional Intelligence at Work | Master Your Emotions," by Six Seconds the Emotional Intelligence Network (course is for purchase)

- **Book**: *Emotional Intelligence,* 25th anniversary edition paperback, by Daniel Goldman

Action Accelerators

Leverage these resources to enhance your knowledge and increase the effectiveness of your actions:

- Forbes.com: "Diversity Is a Fact, Inclusion Is a Choice," by Timothy R. Clark: www.forbes.com/sites/timothyclark/2021/03/17/diversity-is-a-fact-inclusion-is-a-choice/?sh=68e412a6bd1a

- InDiverseCompany.com: "The relevance of social intelligence in the workplace," by Manasi Bharati: https://indiversecompany.com/the-relevance-of-social-intelligence-in-the-workplace

- SHRM.org: "Emotional Intelligence Helps Build Inclusive Workplaces," by Paul Bergeron: www.shrm.org/resourcesandtools/hr-topics/behavioral-competencies/global-and-cultural-effectiveness/pages/emotional-intelligence-brings-out-the-best-in-inclusive-offices.aspx

- HBR.org: "How Sharing Our Stories Builds Inclusion": www.shrm.org/resourcesandtools/hr-topics/behavioral-competencies/global-and-cultural-effectiveness/pages/emotional-intelligence-brings-out-the-best-in-inclusive-offices.aspx

- **Book:** *What's Your Story? A Journal for Everyday Evolution,* by Rebecca Walker and Lily Diamond

Sources Cited

Betterup.com. "What is Self-Awareness and Why is it Important?" https://betterupstaging.ingeniuxondemand.com/en-us/resources/blog/what-is-self-awareness-and-why-is-it-important

Bernardo M. Ferdman and Barbara R. Deane, eds. *Diversity at Work: The Practice of Inclusion.* San Francisco: Josey-Bass, 2014 (pages 128–139)

Connect with Your *Why*, Find Your *Why Now*

"I have discovered in life that there are ways of getting almost anywhere you want to go, if you really want to go."

—Langston Hughes

***

Every year there are a few mandatory trainings to attend in the workplace. With an increased focus on diversity, many companies now conduct trainings on bias, sexual harassment, cultural awareness, and other diversity topics to avoid legal action, demonstrate their commitment to a safe and fair working environment, and foster an inclusive culture. Organizations have their diversity and inclusion *why*. The *why now* may range from maintaining brand reputation and the bottom line in the midst of social movements like #MeToo to increased awareness and real concern for employee well-being or in response to an employee impropriety. Courses are intended to help people from different backgrounds work better together, as well as with clients/customers, and better prepare leaders to support diversity efforts. Attendance tends to be company policy and a requirement for employment with the expectation that everyone understands and agrees to comply. While training is a good thing and makes sound business sense, sometimes the unexpected happens; participants emerge feeling confused, judged, resentful, or angry. These feelings are not uncommon, and there are dozens of reasons for these reactions.

Lawrence, a White male senior manager in his mid-forties, feels that racism and discrimination aren't issues as he personally has never experienced them in the organization. He believes that the training was a total waste of time and that those company resources could have been used to fund his next project. He's baffled that the leadership team would do such a thing. Martha asserts that she is not racist; her nanny is Hispanic, and she is proud to say that she doesn't see color. Racism is someone else's problem but not hers. George thinks that diversity training is for the sole benefit of minorities so that they can get "special treatment" and complains of reverse discrimination. The multicultural messaging makes him feel left out and builds feelings of animosity. Janet is overwhelmed with the increased awareness of her bias, times of insensitivity to racial injustice, and the countless unintended microaggressions she's committed. She now feels as if she is walking on eggshells. Fernando is skeptical of the company's efforts and sees it as a check-the- box exercise. Months later, nothing has changed except that he is expected to be the voice of all Hispanics since he's the only one, where for years, no one ever expressed interest in his opinion or concerns.

What's particularly interesting is that leaders expect these one-and-done trainings to have a lasting impact on company culture as new knowledge and tools have been imparted to encourage new ways of interacting. An article by Harvard University explains, "Hundreds of studies dating back to the 1930s suggest that anti-bias training doesn't reduce bias, alter behavior, or change the workplace. Two-thirds of human resources specialists report that diversity training does not have positive effects, and several field studies have found no effect of diversity training on women's or minorities' careers or on managerial diversity." That said, while training does indeed offer a new perspective and increased awareness of behaviors and attitudes that get in the way of improved working relationships—there is no reason to expect *that* alone will change life-long beliefs and interactions based on those beliefs. That is a conscious choice to be made solely by the individual. We've got to decide what to do with the newly acquired information and choose to practice it every day if we genuinely care about the well-being of others. Start by connecting with your *why* and your *why now*. As a *Forbes* article describes, "If one knows their why, it's a lot easier to anchor to that stated mission, put in the time and make a concerted effort to go after the objective.

D&I (diversity and inclusion) is very much an individual journey where folks need to get introspective, peel back the layers of their experiences to help dismantle the fear, and learn what their respective mental blockers might be." Once you've discovered your *why*, cement it with your *why now*. Why now after years of thinking and behaving in the same way, would you change? When we do things that we believe make a meaningful impact toward our principles and philosophies, it's human nature to feel happier, more fulfilled, and thus, more inclined to apply the knowledge.

Christina, a regional manager at a prominent mobile phone service provider, experienced this transformation firsthand. Training raised her awareness of her biased beliefs that Black people were lazy and not as driven as White people, yet she was still blinded to the exemplary work of Tamara, a single Black mother of two and an employee of four years. Tamara's customer survey scores were regularly above 90 percent, and she consistently ranked in the top three of the seven frontline sales representatives. Tamara was indeed ambitious and worked hard to provide a good life for her sons but was looked over for the opportunities of team lead on one occasion and store manager in another, in favor of White co-workers with lower satisfaction scores and sales numbers. Christina could never provide an objective reason for her decisions except that Tamara wasn't ready yet and needed to be more like the individuals who were promoted. Tamara wondered what that meant. It was during the follow-up discussion, after the second denial for advancement, that Tamara asked for more specifics and examples in an effort to do better next time. As Christina shared her observations about the promoted individuals, she realized that they were very much like her. They came from similar backgrounds and had yet to start families. Her assessment of readiness wasn't based on skills and competencies, which Tamara had clearly demonstrated, but her view that capable people were like her, White, college educated and without children. Now that Christina was aware of how her stereotypical beliefs about Black people and her model of success informed her decisions about Tamara as well as the impact of those decisions—that a very hard-working candidate was blocked from consideration—she knew that she had to change her thought process. She had the *why* and *why now* she needed to look for ways to interrupt her biased thinking and decided to get to get to know Tamara by mentoring her. In the process, she discovered the sacrifices Tamara made on behalf of the company like working off-days to cover the shift of co-workers who called in sick, and she also learned of Tamara's dreams of homeownership. Christina began to feel more fulfilled knowing that her guidance was positioning Tamara for success and the realization of her dreams. Ultimately, Tamara accepted a manager's position in a different industry and continues to communicate with Christina. Christina has become more objective in the evaluation of employee skills for promotion opportunities and continues to learn and give back by volunteering at a local Junior Achievement—a not-for-profit organization that helps prepare students for a successful future.

The knowledge and skills acquired in diversity training wane over time as people return to old habits, especially when there is no company ecosystem in place to encourage and reward inclusive behaviors. We've got to decide whether we sincerely want to make the effort to be more inclusive, what beliefs need to change, and what actions we need to take. Ask ourselves questions that cause us to reflect. Consider starting with "Do I believe in equity for all, and how do my actions reflect what I believe? What can I do to make a lasting difference that goes beyond what I am currently doing?" Whether a training or other vehicle like books, webinars, etc., the application of the knowledge is where you can begin to make an impact. As champions of inclusion, we are continuous learners who seek to evolve our thinking and alter behaviors that create inclusive experiences with others. We endeavor for enhanced perspectives. We disrupt our biased thinking so that we can model what impactful inclusion looks like in the workplace and beyond. The outcome is well worth the effort.

Actions

Get Acquainted with Your *Why*

Connecting to your *why* will require some soul searching as you explore the inner you. Gain clarity by having an open and honest dialogue with folks you trust about your self-discoveries or revelations. Those close to us often know us better than we know ourselves and can help us connect better to our *why* and *why now*.

Keep Your *Why* Top of Mind

Create a personal *why* statement. Understand why inclusion is important and how you want people to feel as a result of interacting with you. Consider who else can benefit from your inclusion efforts beyond your place of work. Keep your *why* statement posted somewhere that you can see it every day. It's your *why* that will keep you forging ahead when you feel like giving up.

Go Beyond One-and-Done Training

Commit to ongoing learning. Create a newsfeed related to diversity topics, various cultures, or legislation impacting underrepresented groups to stay current on the issues and identify areas that align with your *why* for greater impact.

Stay Connected to Your *Why*

Everybody loves to measure their success when they commit to something. Journal about the application of your new knowledge, its impact, and how that makes you feel. Capture mistakes, what you learned, and new connections made.

Action Accelerators

Leverage these resources to enhance your knowledge and increase the effectiveness of your actions:

- YouTube.com: *How to Outsmart Your Own Unconscious Bias*, Valerie Alexander, TEDxPasadena: www.youtube.com/watch?v=GP-cqFLS8Q4&t=905s
- YouTube.com: *We Are Not a Melting Pot*, Michelle Silverthorn: www.youtube.com/watch?v=PnwTnYE_onQ
- Wseap.com: "6 Tips on How to Be Inclusive at Work," Workplace Solutions: www.wseap.com/how-to-be-inclusive-at-work
- HBR.org: "Why Diversity Programs Fail and What Works Better," by Frank Dobbin and Alexandra Kale: hbr.org/2016/07/why-diversity-programs-fail

Sources Cited

Frank Dobbin and Alexandra Kalev. "Why Doesn't Diversity Training Work? The Challenge for Industry and Academia," scholar.harvard.edu/files/dobbin/files/an2018.pdf

Bernard Coleman. *Finding the Why in Diversity and Inclusion*, February 25, 2019, www.forbes.com/sites/forbescoachescouncil/2019/02/25/finding-the-why-in-diversity-and-inclusion/?sh=3d4a957b4f24

Create New Habits

*"We should indeed keep calm in the face of difference and live our lives in
a state of inclusion and wonder at the diversity of humanity."*

—George Takei

***

I have yet to meet anyone who believes that they are just average as they
assess their knowledge and skills. When it comes to acts of inclusion,
it's easy to consider ourselves above average especially when we are not
consciously seeking to be exclusive in our behaviors and interactions.
Though you may believe that you are on top of your inclusion game, my
money says that there's always room for improvement. Psychological
research suggests that we are not very good at evaluating ourselves
accurately. This natural tendency to overestimate our competencies in
disciplines such as money management, reading people's emotions, or
driving is described by researchers as the Dunning-Kruger effect—the
cognitive bias whereby people with low ability at a task overestimate
their ability. A *New York Times* article summarizes that "Dunning and
Kruger's research shows that underperforming individuals reach erro-
neous conclusions and make unfortunate choices, but their incompetence

robs them of the ability to realize it. This incompetence, in turn, leads them to hold inflated views of their performance and ability." What's notably interesting according to the research is that those with the least ability are often most likely to overrate their skills to the greatest extent.

Meet Brad. He's been a production manager at a manufacturing company for 3 years and has 16 direct reports. His organization works to be compliant with all federal workplace harassment and discrimination laws, ensures leaders attend requisite trainings, and is making progress on hiring people of difference. Brad believes himself to be a well-intentioned, ethical, and a supportive leader who openly embraces diversity. Yet, his department has the highest turnover and recently a few staff members have complained to human resources that he has treated them unfairly. In subsequent meetings with human resources, Brad was surprised to learn that his interactions with women of color demonstrated bias, bigotry, and racism that negatively impacted their day-to-day experience within the organization.

Unfortunately, we are all prone to an inflated view of our expertise as we rely on current competencies to be successful, as we don't know what we don't know. The experts argue that when we lack knowledge and skill in particular areas, we make mistakes leading to poor decisions, and those same knowledge gaps blind us to our errors. In other words, when we lack the very expertise needed to recognize how badly we are doing, we continue to perform badly. The Dunning-Kruger effect isn't about ego making us oblivious to our weaknesses, as we tend to acknowledge them once we become aware of their existence. In one study, students who initially underperformed on a logic quiz and then took a mini course on logic were then able to recognize that their original performance was more flawed than they thought. This may be why people with a modest level of skill or experience temper confidence in their abilities as they know enough to know that there's much left to learn. Like anything else, in building skills, we need to consider the constructive feedback of others with an open mind and be on a continuous path of learning from various sources. We may choose to observe others who are getting it right, attend workshops, or read books followed by the application of what we've learned. From that work, we change our actions, and over time these actions develop new habits. Good habits create positive change and eventually become routine, thereby requiring little cognitive energy or effort to execute. Imagine the impact from the habit of greeting people authentically, listening as an ally, or creating a safe space for you and your co-workers to have courageous conversations. As champions of inclusion, we commit to the development of inclusive habits and live by that commitment. We actively mix up our routine to engage with more people and keep insights fresh. Rest assured, there are individuals in our organization who feel watched without ever being seen – signaling "I don't belong here." We can mitigate that with inclusion habits built into everyday interactions. It's about paying closer attention to the feelings of others and no longer assuming or not caring. The hardest part is getting started. Start small if you need to. But start. Create new habits with a purpose.

Check out 10 habits to work toward in Figure 3.1.

1.   Elevate equity.
2.   Gravitate toward difference.
3.   Amplify voices.
4.   Acknowledge inclusive efforts of others.
5.   Self-educate.
6.   Learn from mistakes.
7.   Call in–call out.
8.   Advocate for others.
9.   Encourage others to be inclusive.
10. Create psychological safety.

Figure 3.1: Ten habits to work toward

Actions

Understand How Beliefs Impact Actions

Consider the lives of three colleagues who are different from you. Expand your view beyond race and gender to include age, socio-economic background, disabilities, etc., and reflect on societal norms that impact what you believe and how you treat them. Ask yourself, "What can I start doing to make them feel more included?" and "What can I stop doing that make people feel excluded?" and do those things consistently.

Consider the Impact of Exclusion

Recall a time when you were an outsider and felt excluded. What impact did it have on you? Ponder the emotional toll of that happening to you over and over for years. Next, think through what someone could have done in those moments so that you felt included. Did they need more empathy, compassion, a sense of fairness or courage to speak up? Decide what you need in order to be that person for someone else.

Plan to Act on Your Inclusion Commitment Daily

Now that you have connected to your *why* as outlined in the previous activity, it's time to commit to it. Begin your day by asking yourself, "How can I demonstrate my commitment to diversity, equity, and inclusion?" Write it down and celebrate when you've accomplished it.

Action Accelerators

Leverage these resources to enhance your knowledge and increase the effectiveness of your actions:

- DiversityJournal.com: "The Inclusion Habit," by Amanda J Felkey, PhD: diversityjournal.com/20611-the-inclusion-habit

- FastCompany.com: "Why Showing Up as a 'Comrade' is the First Step to Inclusion in a Remote Environment": www.fastcompany .com/90533099/why-showing-up-as-a-comrade-is-the-first- step-to-inclusion-in-a-remote-environment?partner=rss&utm_ source=rss&utm_medium=feed&utm_campaign=rss+fastcompany &utm_content=rss?cid=search

- LinkedIn.com/pulse: "Everyday Acts of Inclusion," by Victor Dodig: www.linkedin.com/pulse/everyday-acts-inclusion- victor-dodig

- YouTube.com: *Inclusion Revolution*, Daisy Auger Domínguez, TEDxPearlStreet: www.youtube.com/watch?v=u-VMr51yiVc

Sources Cited

Allison He. "The Dunning-Kruger Effect: Why Incompetence Begets Confidence," NYtimes.com, www.nytimes.com/2020/05/07/ learning/the-dunning-kruger-effect-why-incompetence- begets-confidence.html

David Dunning. *Why incompetent people think they're amazing,* YouTube.com, www.youtube.com/watch?v=pOLmD_WVY-E

4

Make the Connection

"I define connection as the energy that exists between people when they feel seen, heard, and valued; when they can give and receive without judgment; and when they derive sustenance and strength from the relationship."

—Brené Brown

***

Every workplace has varying degrees of diversity—from the most obvious of race, age, gender, and culture to the sometimes less obvious of education, experience, socio-economic background, and ability. It's pretty much impossible to get through the day without encountering someone different from you. We often miss out on the rich experiences of getting to know our co-workers in the quest for meaningful work or just getting through the shift. To-do lists, priorities, meetings, deliverables, deadlines, setbacks—you name it, all are part of just getting through the day. Our focus is on *what* we're doing and not on *who* we're doing it with. Since a great portion of our day is consumed at work, one may consider it a no-brainer to develop relationships. By nature, we are social beings. Connecting with one another makes the work more meaningful and more enjoyable and supports the organization's bottom line as good working relationships are paramount in getting the work done. The stronger the relationship, the more comfortable we are in being ourselves, voicing our opinions, and contributing ideas. Good working relationships also provide peace of mind. Consider the emotional toll of working with a colleague that you find difficult. The effort spent in avoiding the individual, strategizing on how to spend as little time as possible in their presence, and lamenting about the messed up situation is energy you can never get back.

When building relationships, especially social ones, naturally we seek people who are like us. We prefer lunch, watercooler chats, and instant messaging with folks around the same age, who share our world views, race, gender, or class, or come from our hometown. In the workplace it is unrealistic to expect that you'll have relationships with only the people who are like you. The process of getting the work done demands working across differences. While you may have figured out how to peacefully coexist with colleagues who are very different from you, that's not the same as connecting and developing interpersonal relationships. It's in this space that relationship development requires more energy as you shift from seeing difference as a problem to valuing it as a strength. True success is an interdependent relationship, and the relationship is the building block of inclusion, which helps us embrace diversity. It's essential that we build relationships where we feel connected to one another as we partner toward a shared goal. The connection matters more than you may think and even more so for our colleagues who work remotely. Connection creates a bond between people when they feel seen and valued. It gives us a sense of belonging in a group, a sense of identity in distinction to others, and a sense of purpose in being a part of something bigger than ourselves. Chanel was hired right out of college to join a team of five sales representatives for a software company. She graduated magna cum laude in mathematics and racked up years of experience working her way through school. She was the only person of color on the team and one of several in a companywide staff of 75. Susan, her manager, believed Chanel's analytical skills and service-oriented background would be an asset to the team as they expanded into new markets. Chanel would bring a much-needed fresh perspective in support of team goals. Susan's expectations might have been met as Chanel proved herself by bringing her A game each day, learning the product inside out, and was exceeding goals by her fourth month; except Susan had not prepared the team to cultivate a relationship with Chanel. She never had to. The team was homogenous and survived on groupthink. A practice described by *Psychology Today* as "a phenomenon that occurs when a group of well-intentioned people makes irrational or non-optimal decisions spurred by the urge to conform or the belief that dissent is impossible. The problematic or premature consensus that is characteristic of groupthink may be fueled by a particular agenda—or it may be due to group members valuing harmony and coherence above critical thought."

Though Chanel worked alongside them, she felt isolated and disconnected. She struggled to fully understand her role and how to contribute to the team's success. While Chanel was highly productive, the team was missing out on her insights, and Susan couldn't figure out why. Chanel approached Susan about her feelings of isolation and not feeling part of the team, and Susan knew something needed to change in order to maximize the team's potential. They devised a plan where the team met weekly to discuss one another's sale opportunities and challenges and explore solutions, and Susan was intentional about making space for Chanel in the conversations. In addition, Susan partnered her with a peer, Brad. His role was to provide context to the team dynamic and shed light on cultural norms and unspoken rules. Over time, the team got to know each other on an interpersonal level. The challenges of one became the challenge of all as they continued to work toward goals. The new structure facilitated connection and camaraderie and positioned them to leverage one another's strengths.

It's important to make a conscious effort to expand your circle and get to know others. Look for opportunities. They're everywhere. For example, during the next team meeting or company event, introduce yourself to folks whom you rarely encounter of a different race, age, gender, or other visible difference. Be genuine. Strive to make the connection. A new point of view awaits. Mentoring or reverse mentoring opportunities may present themselves. The exposure to new people has its advantages. We never know who is going to inspire our next big idea or who we will inspire. Champions of inclusion understand what exclusion can mean to the everyday experiences of co-workers and transform good intentions into meaningful connections.

Actions

Expose Yourself to Difference

When making social plans, choose to attend events that cater to people and interests different from your own, and strive to connect to at least one person.

We tend to seek advice or collaborate with those who are like us. Today, make it a point to seek the point of view of someone who is not like you at all. Strengthen the connection with a "tell me more" attitude.

Connect to the Issues

Learn from those actively involved in driving change. Follow people who are different from you on social media to get a sense of what is going on in communities of difference and the work being done. The exposure will identify movements that you can connect your efforts to. Check out Sybrina Fulton, social change activist; Minda Harts, workplace and equity consultant; and Stacey Abrams, voting rights activist.

Facilitate Connections

Make time to foster and facilitate connections with co-workers on other teams and departments, and don't forget about remote colleagues. In the process we not only create and expand our network, but we also do the same for others. Bring folks together whom you believe have synergies or are working to solve similar problems. Mutually beneficial connections cultivate inclusion.

Adjust Your Lens

Shifting our point of view allows us to see things in a whole new light. Ever notice that when you are too close or too far away from an object that it's difficult to discern what it is and how to respond? If we don't adjust the angles when we are in the unfamiliar territory of fostering inclusion, we make assumptions, repeat patterns, and easily miss opportunities in plain sight.

Action Accelerators

Leverage these resources to enhance your knowledge and increase the effectiveness of your actions:

- ■ YouTube.com: *Building Connections: How to Be a Relationship Ninja*, Rosan Auyeung-Chen, TEDxSFU: www.youtube.com/watch?v=cBmMZFMPf18&t=632s

- YouTube.com: *How to Connect to Anyone*, Soraya Morgan Gutman, TEDxWilmingtonWomen: www.youtube.com/watch?v=4JNElVrSTjs
- TED.com: *How to Connect with People who are Different than You*, Abigail Spanberger, TEDxMidAtlantic: www.ted.com/talks/ abigail_spanberger_how_to_connect_with_people_who_are_dif- ferent_than_you
- TED.com: *10 Ways to a Better Conversation*, Celeste Headlee, TEDxCreativeCoast: www.ted.com/talks/celeste_headlee_10_ ways_to_have_a_better_conversation

INTERVIEW: TELLING IT LIKE IT IS. . .TO GET WHERE WE NEED TO GO

Why Everyone Needs Interpersonal Skills

Meet Khaite. She is a marketing manager at a world-renowned hospital in Illinois. Her high energy and professionalism consistently get her noticed. She has a reputation for going above and beyond in all she does and prides herself on her ability to connect with anyone. As the youngest African American on the team, who happens to be a lesbian, she feels that there is a special lens on her and pressure to represent not just Black people but also the LGBTQ community. Originally hired as a writer reporting to an Asian American woman, everything had gone smoothly for the first two years. When the department restructured, her role expanded. She was given a camera and informed that she was now on the digital team reporting to a middle-aged White man, Richard. He originated from Kansas and had a very 1980's management style. Khaite continued to go above and beyond in her new role and thought that she and Richard were adjusting to one another okay. The relationship was not quite as solid as the one with her previous manager, but after six months of working together, she felt that it was as good as it was going to get. Through seemingly insignificant circumstances, Khaite discovers that Richard is a bit uncomfortable in his new role as her boss.

After an event at the hospital, Khaite hurried back to the office to start her next project. After a few minutes, Eric, a White co-worker sends a text saying that Richard is requesting that she pick up the artwork from the event that he had forgotten and that he needed in time for an event the following day. This wasn't the first time Khaite received requests from Eric on Richard's behalf. Already deeply focused on the new project, she responds that now is not a good time and she will soon be in a meeting. She would be happy to do it afterward except the building where the artwork is located will be locked before the meeting ends. Eric is insistent that Richard wants her to carry out the request. She apologetically reiterates the conflict and takes the time to find someone else to help. She found the situation unsettling as she could not make sense as to why Richard was so unwavering, especially since both he and Eric were still in the area. She assumed that they were just too busy as well. The next day Eric paid her a visit, which wasn't unusual, except this time instead of a big smile and casual attitude, he had a tense look on his face. He began by thanking her for finding a resolution to the artwork dilemma and shared something Khaite will never forget.

A few months earlier, Richard approached Eric to keep a watchful eye on Khaite and to make sure that she stays in her lane. Eric questioned why he needed to become her second and unofficial boss. They shared the same title, he had no more authority than Khaite, and the added responsibility fell outside his job role. Besides, he considered Khaite a friend, and the directive made him feel like a traitor. The situation haunted Eric, and he felt the need to tell her. Richard never made it clear to Eric as to why Khaite needed the extra pair of eyes. But for Khaite, this revealed that despite her best efforts to connect with Richard, she always felt uncomfortable in his presence, and this gave some insight as to why Eric was tasking her with requests from Richard.

Khaite contemplated whether Richard was ill-equipped to lead a diverse team or whether he had some sort of personal problem with her. While the team was gender and age diverse, her presence added a few extra dimensions that may have been unfamiliar. Could it be that she was Black, the youngest, or gay? Could it be that being the "only" on the team, where she did not fit the persona of others who held the title and as a result deemed less competent. Worse, what if he was racist or homophobic? She could not fathom a reasonable explanation as to why Richard did not want to interact with her directly. She pondered whether to approach him but was concerned that it may negatively impact Eric's relationship with Richard or that it may increase the tension with her and Richard. She wondered whether talking to human resources was appropriate, but there were no policies broken that she knew of, so what good would that do? She decided to stick it out and hope for the best. Khaite continues to feel that the relationship with Richard would be much improved if he would take the time to get to know her rather than brush her off as a problem on someone else. The culture created by Richard is now awkward for both Khaite and Eric and created tension in a previously positive and productive work environment.

Experience Other Cultures

"People of different religions and cultures live side by side in almost every part of the world, and most of us have overlapping identities, which unite us with very different groups. We can love what we are, without hating what—and who—we are not. We can thrive in our own tradition, even as we learn from others, and come to respect their teachings."

—Kofi Annan

***

Understanding the unique cultural differences of co-workers can go a long way in creating a welcoming workplace. Things Americans take for granted like the handshake greeting and engaging in eye contact during conversation, for example, may fall outside the tradition of others, and colleagues may come across as standoffish or rude. Shaking with the left hand is considered insulting in the Middle East as the right hand is used for eating while the left hand is reserved for washing after using the bathroom. That said, one must always use their right hand. African Americans are compelled to work longer hours as a means to repudiate stereotypical beliefs of being lazy and staying employed, while for the dominant culture, it's a demonstration of commitment to work and means to advancement. Addressing someone as ma'am or sir demonstrates respect in some cultures and is discouraged in others. By learning and understanding different cultures, we understand why people do things the way they do. The more we know, the better we can connect and communicate. There's no better way to acquire knowledge of other cultures than through experience. The feelings and the lessons learned when experiencing something new can stay with us for a lifetime. Consider your last vacation to somewhere you've never been, especially if it was not your country of origin or country of current residence. Immersing yourself, even for a short time, can change your perspective and help you learn more about yourself. My first vacation to the Bahamas was eye-opening. Coming from the fast-paced environment of Chicago, there are many things I've grown accustomed to, like being serviced in a timely and attentive fashion in restaurants. I remember sitting at the table for what felt like a million years before anyone acknowledged me. I think of how frustrated I became while waiting and vowed to leave an insulting tip. When finally approached by the server, a seemingly middle-aged woman with a big, beautiful smile asked what I would like for breakfast, her smile melted away some of my frustration but didn't stop me from complaining to her about the wait. She continued to smile and didn't address a single gripe. Rather than push for an apology, which I would have received in Chicago, I decided to let it go and work toward enjoying the day. Imagine my temperament when the exact same thing happened again at dinner at a different restaurant. When I returned to the hotel fuming, I vented to a fellow American who frequented there. She explained that the laid-back atmosphere is one of the many things that bring her back again and again. She admitted that it took some getting used to and encouraged me to embrace it. I pondered that whole laid-back thing for a minute. It was in that moment that I realized how impatient I am and how rude I can be when my expectations are not met. This may sound corny, but I was indeed forever changed. For the duration of the vacation, I reminded myself that I was not in a rush for anything, and I should be fully present to truly enjoy the experience of being there. I decided that while in Rome, I would live as the Romans do. It felt great! Returning to Chicago, feeling refreshed, I chose to maintain those lessons learned. I became more observant of the space I'm in before letting emotions take me places where I don't need to be and cause me to react in ways that may harm others. That experience helped me to realize that there are numerous approaches to life, that my approach may not always be the best, and that I should always be curious rather than judgmental when in a new environment.

A cross-cultural experience in the workplace is unavoidable as organizations expand globally and continue to diversify beyond age, race, and gender. As we interact with people from different cultures, we gain a different perspective that can enable us to establish new ways of thinking as well as approaching and solving problems. We get to reconsider our backgrounds and establish where we may have learned culturally insensitive opinions and habits. By acknowledging personal biases and being ready to adjust our views, we start appreciating people from diverse groups for who they are. When we are intentional about truly connecting cross culturally, we expand our knowledge of their history, traditions, values, and beliefs. Subsequently, cultural competence increases, and we can begin to bridge the gap. Get to know the history and significance of cultural celebrations of your colleagues, which may include National Hispanic Heritage Month, Lunar New Year, or Kwanzaa. Discovering the *"why"* behind the traditions not only offers a better appreciation for our differences but provides a more well-rounded approach to the way we engage. There is so much beauty and richness coming from all cultures. Champions of inclusion work to acquire an understanding of appropriate behaviors for better interactions with co-workers of different cultures and build competencies that demonstrate respect. Take the time to plan your work and work your plan.

Figure 5.1 shows a variety of people who may be in your workplace.

Figure 5.1: Individuals representative of various cultures and religions

Actions

Put Yourself in Spaces to Experience Various Cultures

Take a break from the ordinary. Explore fun ways to immerse yourself in new environments with people who are different than you, and discover a whole new world. Here are a few things to try:

- Taking a multicultural cooking class
- Visiting multicultural museums, art galleries, and local fairs
- Volunteering for organizations that serve underrepresented groups or communities
- Traveling with purpose—set aside time to detour from the tourists' spots and mingle with the locals, engage in conversation, and return with a fresh perspective

Educate Yourself on the Cultures of Friends and Colleagues

The more we understand about an individual, the easier it is to build a relationship and work together. When you examine the impact of cultural differences on interactions between individuals nurtured in different cultures, for example, a Chinese manager leading a team of Americans in the United States, you have to consider to what degree, and in what way, do their values differ from Americans in trust building, communication, and teamwork. Research the unique background of the various cultures that exist in your workplace to increase your understanding. When you work to understand, you open the door to being understood.

Make Curiosity the Default

Consistent interaction with individuals different from you is destined to create awkward moments when the unexpected happens. Instead of ignoring, judging, or assuming, ask questions. During conversation, express interest in learning more about your colleague's culture. Be genuine. Have an open mind. Share stories about your culture, traditions, and beliefs. Do more listening than talking.

Action Accelerators

- YouTube.com: *Cultural Diversity: The Sum of Our Parts*, Hilda Mwangi, TEDxUCSD: www.youtube.com/watch?v=7tv7NaV47no&t=385s

- YouTube.com: *How Culture Drives Behaviours*, Julien S. Bourrelle, TEDxTrondheim: www.youtube.com/watch?v=l-Yy6poJ2zs

- HBR.org: "3 Ways to Improve Your Cultural Fluency," by Jane Hyun and Douglas Conant: hbr.org/2019/04/3-ways-to-improve-your-cultural-fluency

- **Book**: *Your Unique Cultural Lens: A Guide to Cultural Competence* by Prof Enrique J. Zaldivar

Remove the Labels

"Definitions belong to the definers, not the defined."

—**Toni Morrison**

***

We have labels for most everything and everybody. Labeling makes life easier as it defines our expectations and helps us compartmentalize situations and behaviors. The dictionary defines labels as a "classifying phrase or name applied to a person or thing, especially one that is inaccurate or restrictive." You may encounter this more often than you think. Imagine that you've been taking notes with a black pen that's just run out of ink. You find a box labeled "Black Pens" and naturally expect to find black pens inside. But what if instead, you expected black pens with a felt tip and that's not what was in the box? Then that "Black Pen" label did not go far enough in helping you understand its contents. The pen loses some of its identity in the label. In the same vein, when we label individuals, we can easily overlook how unique they are and miss out on their richness. When we encounter someone with an ethnic or non-American sounding last name, with brown skin tones, or speaking with an accent, we're inclined to believe that we know their lineage which limits our curiosity about them. The fact is that we can't judge a book by its cover. In a globalized world, I am meeting more individuals who are multiracial, multicultural, or a combination thereof, and cannot and should not be categorized as "this race" or "that culture" lest we lose a big part of their identity. Labeling puts us in the frame of mind that leads us to believe that we know everything we need to know about someone often before they even speak. We're wired to put people in our boxes of expectations and engage within those boundaries. As we strive to be more inclusive, we must divert from a one-size-fits-all mindset to seeing the whole person and their uniqueness. Tiger Woods, a renowned American golfer of African American and Thai descent, refuses to identify as Black as he embraces his mixed-raced heritage. He sees himself as Cablinasian—a mix of Caucasian, Black, American Indian, and Asian. It's a term that he created with precision as he is one-fourth Black, one-fourth Thai, one-fourth Chinese, one-eighth White, and one-eighth American Indian. When the world looks at Tiger and sees only a Black man, we have in effect written his mother out of his racial identity. If one had to check a single box for Tiger's race, it would be labeled "other," while Tiger may choose to check all boxes that apply. Racial and cultural identity can be an individual choice for many. Barack Obama is biracial. He proudly stated in his 2008 campaign speech in Philadelphia, "I am the son of a Black man from Kenya and a White woman from Kansas." He was raised by his White mother and White grandparents and will forever be remembered as the first Black president of the United States. He identifies as Black, married a Black woman, and raised two Black daughters.

Society as well as the government has a practice of categorizing individuals by descent and bestowing labels that are not universally embraced by the community that has been labeled. As facets of diversity continue to expand in the workplace, more and more people want to be referred to by terms that they have chosen and better align with their history and heritage rather than the labels selected by society. Consider how your perception may change when you see beyond an African American, the lens of which we view Black people, to Jamaican. That distinction can add depth that was not there before. We tend to refer to our Spanish-speaking colleagues as Hispanic, where they may prefer Mexican or Puerto Rican. Our Asian colleagues may prefer Japanese or Korean, while our Middle Eastern colleagues may prefer Lebanese. An increasing number of people are embracing their full identities and we should too. Too often, we assume instead of taking the time to get to know people and let them share on their terms or worse, asking in such a way that offends. While some colleagues are fine with the labels, others feel erased. Taking interest in how people see themselves demonstrates sensitivity and respect. When we understand how individuals see themselves, it makes room for deeper conversation and appreciation for cultural nuances. Make someone's day by getting to know them as an individual, rather than part of a group. Champions of inclusion accept other's references and identities of themselves as truth.

Actions

Expand Your View of People Beyond Race

Shift your thinking from interacting with people based on what you believe their race to be or labels that society has bestowed. Rather, explore what lies beneath. Far easier said than done, I know. Our inclusion journey requires us to go deeper and consider positive things that different races and cultures bring to enrich our lives.

Test Your Assumptions

Consider the last breakfast cereal, big pharma, or life insurance commercial you saw on television. I find that these commercials will often depict a variety of people. List what you observed about the main characters. Assess to what extent you believe your assumptions or labels to be true in three categories: very accurate, accurate, and not sure. Consider how your confidence in your assessment would impact how you engage with them. Create a list that people may make about you. Does it mirror regional or national expectations? Are pieces of your identity overlooked that may skew their perspective? Reflect on how that oversight makes you feel and how it may drive interactions with you.

Discern Whether Your Curiosity Comes from a Good Place

Sometimes people get offended when asked about their race or culture. Often, it's due to the frequency with which the question is asked and when it's not coming from a good place. Perhaps it comes across as intrusive and feels like a judgment question, especially when you first meet someone and have not established a rapport. Before posing the question, consider whether you really need to know. Are you asking because you want to make sure that you're getting it right, or are you unconsciously othering them? *Othering* is defined by VeryWellMind.com as "a phenomenon in which some individuals or groups are defined and labeled as not fitting in within the norms of a social group. It is an effect that influences how people perceive and treat those who are viewed as being part of the in-group versus those who are seen as being part of the out-group." Will knowing change or improve the relationship? When asking give context as to why you're asking.

Action Accelerators

- Vox.com: "The inadequacy of the term 'Asian American,'" by Li Zhouli: www.vox.com/identities/22380197/asian-american-pacific-islander-aapi-heritage-anti-asian-hate-attacks
- CBSnews.com: "Not all black people are African American. Here's the difference," by Cydney Adams: www.cbsnews.com/news/not-all-black-people-are-african-american-what-is-the-difference
- YouTube.com: *What we lose when we let languages and cultures die*, Bruno Beidacki, TEDxKentState: www.youtube.com/watch?v=G2xgZkvVS-A
- YouTube.com: *The Dangers of Othering in the Quest to Belong*, W. Kay Wilson, TEDxColumbus: www.youtube.com/watch?v=wFQiS3TUsA4

Sources Cited

AP News. "Tiger Woods describes himself as 'Cablinasian'," April 22, 1997, apnews.com/article/458b7710858579281e0f1b73be0da618

Grant Rindner. "Who Were Barack Obama's Parents?" November 16, 2020, www.oprahdaily.com/life/a34670592/barack-obamas-parents

Widen Your Perspective

"We seldom realize, for example, that our most private thoughts and emotions are not actually our own. For we think in terms of languages and images which we did not invent, but which we were given to us by our society."

—**Alan Watts**

***

We are living in an unprecedented time of profound and positive shifts in attitudes toward race and culture with many of us being awakened for the very first time to the deeply embedded racist views and systems that have existed for centuries. The overwhelming reaction to the murder of George Floyd, the support for Prince Harry and wife Meghan over concerns from the royal family about how dark their baby's skin tone would be, or the uproar in response to states banning critical race theory in education are on the world's stage for all to see. This shift in perspective is long overdue, and we can no longer claim that we didn't know. Most of us alive today are not responsible for the creation of the systems and institutions that have disenfranchised so many; however, we all have a responsibility to do something about how they continue to impact our lives today.

As the workforce becomes more diverse, it's important to widen our perspectives in order to change our frame of reference in relationship to others who are different from us. Our eyes open to a new point of view as we question the validity of belief systems and ideologies that cause harm. Systems and institutions have built a toxic culture based on the conviction that one race is superior to another, that members of the LGBTQ community do not have the right to marry, that racial profiling in policing reduces crime, or that women don't have the right to choose regarding birth control and abortion. Our culture informs the way we communicate and engage with one another. When the video of Amy Cooper leveraging her White privilege and fabricating a life-threatening situation went viral, she put her convictions and beliefs on display. She altered her voice while calling the police claiming that a Black man was threatening her life. In reality, the alleged threat, Christian Cooper (no relation), was birdwatching in Central Park and had simply asked her to place her dog on a leash as required by park regulations. I can't help but wonder why she felt comfortable in her actions at that moment. Did she have the slightest inkling of the potential harm? Her behavior that day is representative of historical beliefs of White entitlement and racism that could have escalated and ended Christian's life at the hands of police for birdwatching while Black and asking a White woman to obey the law. Thankfully, Christian recorded the encounter and posted it on social media. Without Christian's documentation, the truth may have never been revealed, and Amy may have never suffered the consequences of her actions. Her employer fired her citing that they do not tolerate racism. In addition, she was charged for falsely reporting an incident in the third degree. Charges were later dropped upon completion of counseling designed to educate her on the harm of her actions. Her unrestrained convictions cost her dearly. My guess is that she will carry the weight of her decisions for years to come.

Many of us go through life without ever questioning why we say, do, or believe certain things, and those ingrained values can follow us right into the workplace. We continue with unsavory behaviors because that's the way we've always done it or seen it done. Unchecked, the realization that we are offending or harming others and perpetuating unfair systems will never dawn on us. As we build awareness of how culture impacts the ways in which we communicate and how communications demonstrate values, we can move from holding each other back to lifting one another to a state of inclusion and belonging.

As culture is ever-changing, we must ensure that interactions are culturally appropriate. It's natural not to understand how individual communities experience the world, but that's no excuse to not learn and become more aware. Using an open mindset, we can acquire new skills enabling us to communicate more effectively with one another. Having been in the workforce for almost four decades, much has changed, and there's so much more to do. In my early career as a single mother, I was judged and subtly punished for not having a husband, taking a sick day when my daughter was ill, and for carrying a purse *and* a briefcase. Cultural norms of that time were primarily based on Christian beliefs that frowned upon single mothers and childbirth outside marriage and were intolerant of women in the workplace. Women were expected to stay home caring for the children while husbands went to work. The convictions of my bosses and co-workers shamed me and made me feel inferior for not living up to their beliefs. Thankfully, the way it was is no longer the way it is now in the workplace. While working mothers still struggle with being taken seriously and continue to face adversity in the workplace, new cultural norms demand that co-workers and leaders communicate in ways that are more supportive and keep demeaning thoughts to themselves. When we value working mothers and other underrepresented groups, instead of marginalizing them, the narrative changes. Acknowledging that the person standing in front of us or sitting across from us is no less valuable because of difference requires that we see them as equals. We begin to show appreciation for a new point of view, demonstrate respect for another's uniqueness, and are open to course correcting when we get it wrong. Over time, we build a repertoire of what's acceptable and what is not for various demographics to include race and ethnicity, sexual orientation, age, gender, and disability status. Competence with one group doesn't translate into competence with another. As we engage, we discover that we are learning something new almost every day. It's a lifelong journey, and there are no shortcuts. It's well worth your time to take stock of your personal convictions. Challenge the presumption that women with children aren't ambitious. Don't assume that marriage is only between a male and female. It's no longer acceptable to refer to Black people with racist slurs or using the "B" word when referring to women. There is not a single circumstance that I can think of where that would be okay—yet it still happens. The ABC television network cancelled the *Roseanne* show over a racist tweet from star Roseanne Barr that compared former White House advisor Valerie Jarrett, who is Black, to *Planet of the Apes*. Queen bees, women in authority or power who intentionally mistreat subordinate females simply because of gender, should examine their motivations. We can no longer accept double standards rather we must question them. Everyone needs to know more about each other to get to a place of acceptance. Expose yourself to different social environments and self-educate. Over time, these varied experiences will awaken you to harmful, inaccurate, or misleading convictions. Seeing the world with an enhanced perspective is the way forward. As champions of inclusion, we seek and listen to multiple voices and perspectives as we learn to approach difference in ways that support mutual learning and growth.

Actions

Revisit Your Upbringing

Learn about yourself. Explore the root of your convictions. Understand where your values and beliefs originate and check whether they are harmful to others. Once identified, devise a plan for change and put it to use.

Learn from Mistakes

There's a saying that hindsight is 20/20. When we learn from mistakes, we get 20/20 foresight. The only mistakes made are the ones we don't learn from. Incorporate lessons learned into future interactions.

Get Familiar with Thought Leaders of Difference

Changing perspective requires exposure to new ideas and people. Increase your appreciation for the successes of those different from you. Consider Jerry Wang, Taiwanese co-founder of Yahoo; Mellody Hobson, African American CEO of Ariel Investments; Daymond John, dyslexic and African American businessman and investor on Shark Tank; Tim Cook, gay CEO of Apple; and Martine Rothblatt, transgender woman and CEO of United Therapeutics.

Action Accelerators

- YouTube.com: *How Culture Drives Behaviours*, Julien S Bourrelle, TEDxTrondheim: www.youtube.com/watch?v=Tz1LWIqbHas
- PsychologyToday.com: "The Psychology Behind Racism, attempts to understand the inexplicable," by Allison Abrams:

www.psychologytoday.com/us/blog/nurturing-self-compassion/
201709/the-psychology-behind-racism

■ OpenText.wsu.edu: *Module 5: Attitudes*, Press Books: opentext
.wsu.edu/social-psychology/chapter/module-5-attitudes

■ Inclusion.AmericanImmigrationCouncil.org: "Individual Change
or Societal Change. Which Ensures Durable Shifts in Attitudes and
Behaviors?" by Wendy Feliz: inclusion.americanimmigration
council.org/durable-changes-in-attitudes-towards-
marginalized-groups

8

Slow Down Before You React

"People will forget what you said, people will forget what you did, but people will never forget how you made them feel."

—**Maya Angelou**

***

Ooops!! You've been called out for a microaggression—a more often than not unintended insult or slight that expresses a prejudiced attitude or stereotypical belief toward a member of a marginalized group. This is difficult to self-identify because we are unaware that it's happening. Sometimes they happen without having to utter a word. Consider whether you've clutched your purse, wallet, or briefcase tighter because a Black man joined you in an elevator or you quickly shift from a watchful eye to closely following Latino customers around your store or you consistently choose not to sit next to a person of color at conferences or events. Verbal microaggressions are usually intended to be complementary. The intent isn't what matters; it's the impact. Telling a Black co-worker "You speak so well" suggests that you expect them not to be as articulate as White colleagues. Asserting as a White woman to a woman of color "I know exactly what you're going through" signifies that her gender oppression is no different than the other woman's race and gender oppression. Then there are those microaggressions that reveal your biased assumptions. Asking the Hispanic executive at a conference to refill your water glass implies that people of color are servants to White people and couldn't possibly occupy high status positions. Expressing surprise when your tall African American colleague didn't attend college on a basketball scholarship but an academic one suggests that people of color are not as intelligent as White people. These seemingly harmless insults have a cumulative effect and are coined by psychologists as "death by a thousand cuts." A post by *Psychology Today* found clear evidence that daily experiences of racial microaggressions harm the psychological and physical well-being of minorities. Further, data indicate that racial microaggressions are linked to low self-esteem, increased stress levels, anxiety, depression, and suicidal thoughts. While death by a thousand cuts is an accurate assessment, I have to add that some cuts go deep and may feel more like a stab. I recall an instance where I shared with a White colleague that I was considering moving from the city to a particular suburb. As it turns out, it was the town where he was currently residing. I was expecting a response something like "How nice. We will be neighbors." Instead, his response was "Oh yeah, I've seen a few Black people around lately. I wonder how they can afford it."

When a co-worker pulls you aside, shares how your actions made them feel, and asks what you meant by your act or comment, how do you respond? Instinctually, you may become defensive or attempt to explain it away. After all, you don't know what a microaggression is, have never heard the word, and would never in a million years intentionally offend someone. In oblivion, you're shocked that it's been called to your attention. Interrupt that common reaction. It's useless. The fact is that someone was offended, invalidated, dismissed, or otherwise harmed by what you said or did. When we realize that we've hurt someone, as unintentional as it may have been, the common response is to apologize if you value the relationship. Reflect on the last time you accidentally hurt a close friend or loved one. My money says that you immediately apologized and wanted to know what to do to make it right. When you value the relationship with co-workers, the reaction should be no different. Productive conversations happen only when the offender responds to the feedback respectfully, empathetically, and with a willingness to learn. Your co-worker is gifting you an opportunity of personal self-reflection and valuable insight on why your act was hurtful. As with any gift, it's yours to keep or not. Choose to keep it, and it's a gift that keeps on giving through improved interactions and a better working relationship going forward. Genuinely listen, ask clarifying questions, reflect, and take responsibility. Don't forget to apologize. The key to success is being humble enough to listen to the impacted person and understand their perspective. Every journey begins with the first step, and this first step may be uncomfortable as you demonstrate that you've learned something from the experience.

When Sally first joined the team, Jamal was excited to work with her. He admired her analytical skills and can-do attitude, which complemented his big picture thinking. Sally lives in Texas and works remotely. Jamal has been a Chicagoan his entire life and is based at company headquarters. Struggles soon began with her southern hospitality style of communicating. He resented her every time she called him "sweetie" or said "cotton picking" when she disapproved of something. Her contrasting world views often seeped into conversation along with her quest for knowledge on how to be politically correct with Black people. He had to give her credit. At least she was trying to understand, which was more than he could say about other co-workers. He often wondered whether he was the only Black person she knew. Jamal believed that she was genuinely unaware of how she came across and how uncomfortable she was making him. He doubted that she meant any harm and felt that she was a nice person. So, he continued to give her grace and kept quiet about his angst for months hoping things would improve. It didn't. It got so bad that he dreaded their weekly calls and found himself stressed for days leading up to their meetings. As an intervention, he decided to start communicating with her more via email and Slack to get the work done and avoided her "I have a quick question" calls. The effort it took to avoid her was just as frustrating as collaborating with her directly. Feeling hopeless, he decided to have a talk with her. Concerned of the potential that the conversation would not go as he hoped, he began by sharing how much he valued her skills and respected her efforts to get to know him as a person. He continued with how calling him "sweetie" made him feel disrespected and that "cotton picking" was offensive as it wreaked of the slavery of his ancestors. To his surprise, Sally immediately apologized and thanked him for the feedback. She agreed to choose her words more wisely in the future, avoid inserting her personal views into conversations, and invited Jamal to let her know the next time she "put foot in mouth."

Microaggressions can be committed by anyone. They point out differences and magnify them to the extent that the recipient is left with a barrage of negative feelings that eventually have to be reconciled in one way or another. Don't underestimate the impact; work to understand it.

As champions of inclusion, we work through and learn from mistakes as well as listen to co-workers until they feel understood.

Actions

Look for Ways to Identify Your Microaggressions

As microaggressions are often unintentional, we are unaware when we've committed them. Ask someone you trust to alert you (in private) when you've made a comment based on bias or stereotypical beliefs. The feedback will increase your awareness of these beliefs and how often they show up.

Understand the Stereotypical Beliefs of Groups to Which You Belong and How They Make You Feel

Stereotypical beliefs are often untrue. To get an idea of their validity, consider all the groups to which you belong. Include race, gender, age, socio-economic background, political affiliation, religion, sexual orientation, and any other group that comes to mind. List the stereotypes associated with those groups. Do these attributes apply to you? Are they accurate? How do those beliefs make you feel?

Raise Your Awareness of the Impact of Microaggressions

Learning about microaggressions in all its forms is the best way to avoid them. Research the different ways microaggressions manifest.

- *Microinsults*, as defined by Dr. Derald Wing Sue, are subtle acts that convey contempt and disrespect for someone. They are subtle, spur-of-the-moment, and back-handed, and it can be hard to tell if the perpetrator is even aware that what they're doing is insulting. An example would be a heterosexual calling someone gay to imply that they are stupid.

- *Microinvalidations*, as defined by Dr. Derald Wing Sue, are when someone tells or implies to someone that their experiences of discrimination aren't real. These microaggressions aren't just covert, but completely hidden. They often don't even mention the identity they are targeting. The perpetrator almost never realizes they're doing something harmful, doesn't mean to do harm, yet harms nonetheless. An example would be when someone tells you that if you were nicer, more respectable, or more polite, then people wouldn't discriminate against you.

■ *Microassaults*, as defined Dr. Derald Wing Sue, are usually deliberate and on purpose. They can be subtle, but usually aren't. They usually happen when the perpetrator is anonymous, they are being supported by peers around them, and/or they know they can get away with it. There's no guesswork in determining if you were the victim of a microassault. An example would be when an able-bodied person takes or knocks a cane out of the hands of a person who walks with a cane.

As awareness increases, work toward eliminating them in interactions.

Action Accelerators

■ **YouTube**: *Episode 7 – Microaggressions*, Diverse City: www.youtube
.com/watch?v=tXO1VKmBWgg

■ **HBR.org**: "You've Been Called Out for a Microaggression. What Do You Do?" by Rebecca Knight: hbr.org/2020/07/youve-been-called-out-for-a-microaggression-what-do-you-do

■ **Book**: *35 Dumb Things Well-Intended People Say: Surprising Things We Say That Widen the Diversity Gap*, by Maura Cullen

■ **Podcast**: "Episode 34, Let's Talk. . . .Microaggressions," *Cornell University's Inclusive Excellence Podcast*: soundcloud.com/iepod
cast/episode-34

■ **ScientificAmerican.com**: "Microaggressions: Death by a Thousand Cuts," by Dr. Derald Wing Sue: www.scientificamerican.com/
article/microaggressions-death-by-a-thousand-cuts

Sources Cited

Abigail Fagan, reviewer. *Psychology Today*. "The Detrimental Effects of Microaggressions," October 5, 2021, www.psychologytoday
.com/us/blog/evidence-based-living/202110/the-detrimental-effects-microaggressions

SailingRoughWaters.com. http://sailingroughwaters.com/define/
microaggression/types/assaults

Microassaults and Gender Pronouns

Meet Alana. In her early 20s she enjoyed working as a contract employee and trying odd jobs as a way to determine a career that supported her middle-class lifestyle. As if this feat weren't difficult enough, it was compounded by her budding sexuality. Now with a short haircut, preference for masculine attire, and the same skinny body she had as an elementary school kid, she came to the realization that her pronouns were *they/them*. Suddenly, the odds grew slimmer of finding a place of work that Alana could enjoy, would pay them an acceptable wage, and would embrace their identity. The thought of having to hide parts of their identity just to get hired, let alone fit in, was overwhelming. Regardless of the challenges, bills had to be paid, and Alana was determined to muscle through. Their mom always said, "if you don't try, you're guaranteed to fail." Those words kept them going.

Contract work was tolerable for the most part and paid the bulk of expenses, but odd jobs were required to make ends meet and have fun money. Alana wasn't a fan of odd jobs. The dynamic was completely different than that of the contract work. It was far less structured. Alana has a strong work ethic, always did more than was expected, and believed that they would be treated kindly if they were kind to others. Building friendly working relationships and finding co-workers Alana could trust seemed impossible. Most workplaces had high turnover. Staff came and went without concern from management as to why, and managers were more focused on their own agendas rather than the employee experience. After the first day on the job, there would be no contact from managers unless there was a problem. Staff was micromanaged by so-called leaders who were actually peers with limited authority. Further, staff was acknowledged only when they made mistakes while co-workers sought brownie points by throwing one another under the bus. Alana felt alone and disconnected. One of their odd jobs was across the street from a family-owned coffee shop. Alana soon became a regular customer. The staff was friendly, seemed close-knit, and always greeted them by name. During a visit, the barista shared that they were hiring and invited Alana to apply. Weeks later, Alana was on the team. Things seemed to be going good at first. Teammates were nice and helpful as Alana learned the ropes. However, at times, Alana felt uncomfortable. There was a feeling of being watched. When they entered the room, conversations stopped. Alana soon realized what was once believed to be a close-knit team was in fact cliques. It smacked of the in-group, out-group drama of high school days, and interactions were often short and abrupt. Alana tried to ignore the feelings and chocked it up as overreacting and decided to have an open mind. Two months in, the feelings of something amiss could no longer be ignored. While cleaning tables, Alana overheard one co-worker ask another to take out the trash. The reply was "have *it* do it" and pointed to Alana. The two laughed about the "it" comment and then proceeded to ask Alana to take out the trash. An avalanche of emotions rushed in. Alana wasn't sure whether the pain was disgust, betrayal, or utter disrespect. One thing for sure, it wasn't to be tolerated, and there was no way in hell they would allow it to happen a second time. Alana took out the trash as requested and resigned the next day. Management never asked why, and Alana didn't volunteer the reason.

The experience however had a lingering effect. Alana continues to wonder "What if I stood up for myself?" Would management have cared, and would things have gotten better or worse? Is it possible that they could have spared the next LGBTQ staff member from the same pain had management been called out? Alana may never know and is troubled that she missed an opportunity to be an instigator of change.

Respect the Beliefs of Others

"An inclusive world starts with each of us choosing to respect perspectives other than our own, treating everyone with respect and choosing to stand up for others who need our support. More than anything else, this is what going beyond diversity truly means."

—**Rohit Bhargava**

***

Chances are you have co-workers who ascribe to organized religion as well as ones who consider themselves atheists, agnostic, or nonreligious. The Religious Landscape Study by Pew Research Center surveyed more than 35,000 Americans from all 50 states about their affiliations, beliefs, and practices. The study revealed that almost 71 percent self-identified as Christian, almost 6 percent belonged to other faiths, and roughly 23 percent were nonreligious. I find that people who ascribe to a particular religion or spirituality are usually uninterested, judgmental, or even hostile toward other religions, and sometimes justifiably so. The extremist Muslim terrorists' attacks on September 11, 2001, will forever be a part of history. Consider the numerous child sex abuse allegations in the Catholic church, the scandals of Evangelicals like Peter Popoff, a debunked faith healer, or the ABC News investigation revealing how Robert Tilton's ministry discarded prayer requests without reading them, keeping only the money or valuables sent by viewers. Christianity isn't alone when it comes to conflicts of what's espoused and what's actually practiced. In the midst of the #MeToo movement, Nouman Ali Khan, a rigidly conservative celebrity American Muslim preacher, was caught in a sexting scandal; and when one of the best-known faces of American Buddhism, nun Pema Chodron, stepped down as a Shambhala teacher, she alleged sexual abuse and misconduct of its leaders. The malicious acts of a few can raise strong convictions and polarize a workplace as well as invite prejudice and hatred. The onus is on us to not judge all for the acts of a few. When hate crimes are committed based on religious beliefs, co-workers who are of the same faith as the perpetrators are usually just as outraged as everyone else. Being inclusive means getting to understand their point of view rather than assuming that you already know. Misinterpretation, jokes, and judgment create a hostile work environment when we do not understand value systems and the reasoning around religious traditions like prescribed dress, dietary guidelines, and various forms of lifestyle. You may work with individuals who, based on their beliefs, don't celebrate birthdays or Easter and Christmas, object to vaccinations, rely on their God for healing rather than seek medical help, or consider it sinful to intentionally conceal their religious inscriptions.

What I find particularly interesting is how little curiosity there is around affiliations different from one's own. Few of us ever ask folks why they believe what they believe and how those beliefs impact their lives. It does not mean that we must agree, but at least we gain a richer understanding of their personal motivations. As you learn of colleagues who are of a faith different than your own, get to know more about their beliefs. Approach with a curious mindset to avoid innocent mistakes that can offend.

Dalia Mogahed, opinion contributor, wrote in this *USA Today* post, "As an American woman who is visibly Muslim, I can personally attest to the wide prevalence of this perception [that Muslims have outdated views of women] especially on the part of other women. Even among many liberals, the unquestioned assumption seems to be that I am deserving of their pity before their respect. A well-meaning woman approached me recently in a public bathroom to inform me that I was now 'in America' (what?) and that I didn't have to wear that thing on my head here.

A more creative microaggression came from a White woman sitting next to me at a coffee shop. Seemingly out of nowhere, she declared to an adolescent girl, who I presumed was her daughter, that she would never be subjugated to any religion that tells women they are inferior and have to wear the veil. The girl looked mortified. I hope the girl will also never be subjugated to being totally embarrassed in a public place again so her mom can feel superior to a Muslim woman she knows nothing about. Had she taken the time to ask, rather than assume to know me, she might have come to learn that I head research at a DC think tank, I'm an engineer by training, I went to a business school (where I was one of 20 women in a class of 150), and that I was appointed as an adviser to an American president. Also, I find great meaning and joy in my faith and choose to practice hijab as an act of religious devotion—plus, I find it empowering."

Christianity is the norm in the United States, and Christian beliefs manifest in organizational structure to accommodate days like Ash Wednesday, where staff is allowed to arrive later in the day to go and receive ashes on their foreheads, and Good Friday, where the organization is closed the Friday before Easter. As leaders become more sensitive to nonreligious staff, such celebration titles have been renamed or no longer promoted. Good Friday is now referred to as Spring Holiday at many organizations, and later arrivals for Ash Wednesday are supported by encouraging staff to use paid-time-off benefits rather than simply choosing to arrive late. The privilege of late arrival on company time smacks of exclusion to those who don't observe Ash Wednesday. Beliefs express themselves in actions and behaviors in the workplace and may show themselves in a variety of ways from a well-intentioned "Merry Christmas," which Jewish or Muslim co-workers would find offensive, to displays of intolerance for team members of the LGBTQ community or judgment of those who consume alcohol. We must be careful to never express contempt or say anything negative to co-workers about their religion, or lack thereof, or force an unwilling co-worker to listen to our religious views.

Be mindful that not all colleagues share your beliefs and traditions. We owe it to ourselves and them to be knowledgeable and respectful of religious and nonreligious beliefs. Champions of inclusion demonstrate empathy and understanding for what others value as well as believe, and strive to be nonjudgmental.

Actions

Find Common Ground

Becoming more familiar with the values and customs of various religions will enhance our perspectives about other faiths. Educate yourself on different belief systems. You may discover commonalities with your own.

- Attend worship services with a trusted colleague of another religion. Take a learning mindset.
- Create a list of movies to enjoy. Consider:

  - *Arranged* (2007)—A drama about the friendship between an Orthodox Jewish woman and a Muslim woman
  - *Ida* (2013)—A drama about a young nun who is about to take her vows uncovers a family secret dating back to the German occupation
  - *The Passion of the Christ* (2004)—A drama depicting the final 12 hours in the life of Jesus of Nazareth, on the day of his crucifixion in Jerusalem
  - *The Burmese Harp* (1985)—A drama about a Japanese soldier separated from his platoon who disguises himself as a Buddhist monk and begins a journey toward peace of mind amid the chaos

Understand the *Why* of Holy Day Observances

People practice religion for various reasons. A great way to gain perspective into the *why* of a religion can be found in the observances of their holy days. Check out the list in Figure 9.1.

Break Bread Together

Eye-opening conversations can happen over a great meal. Work with your human resources department to host a voluntary lunch-and-learn gathering where co-workers of various faiths prepare a dish relative to their faith and share details about their religion.

Jewish	Christian
Rosh Hashanah	Easter
Passover	Christmas
Yom Kippur	Ash Wednesday
Hanukkah	**Hindu**
Islam	Ganesh Chaturthi
Hijra New Year	Nav Varsh Samvat
Ashura	Vaisakhi
Hajj	**Buddhism**
African American	Mahayana New Year
Kwanzaa	Nirvana Day

Figure 9.1: List of religious holy days and observances

Action Accelerators

- **YouTube.com**: *Finding and Removing the Bias Within Us*, Anthony Wermers, TEDxYouth@Conejo: www.youtube.com/watch?v=i6lRmEdP4fA&t=226s

- **WikiHow.com**: "How to Appreciate People of Other Religions": www.wikihow.com/Appreciate-People-of-Other-Religions

- **YouTube.com**: *Everyday Example: Religion in the Workplace*, Church Newsroom: www.youtube.com/watch?v=xW8lQb0wJZY

- **Book**: *Appreciating All Religions: Religious Literacy in small bites*, paperback, by Paramjit Singh Sachdeva

- **Website**: LearnReligions.com: www.learnreligions.com

Sources Cited

Religious Landscape Study. www.pewforum.org/religious-landscape-study

BBC.com. "Catholic Church child sexual abuse scandal," October 2021, www.bbc.com/news/world-44209971

Darren Cunningham. "Peter Popoff: Miracles or Manipulation?" Fox17online.com, www.fox17online.com/2018/01/05/miracle-or-manipulation

ChicagoTribune.com, "Rev. Robert Tilton, for the moment subdued," March 29, 1992, www.chicagotribune.com/news/ct-xpm-1992-03-29-9201280825-story.html

NPR.org. "Muslim Communities Divided Over Abuse Allegations Against Popular Preacher," *All Things Considered* podcast, November 22, 2020, www.npr.org/2020/11/22/937804223/muslim-communities-divided-over-abuse-allegations-against-popular-preacher

Michelle Boorstein. "Famed Buddhist nun Pema Chodron retires, cites handling of sexual misconduct allegations against her group's leader, Washington Post.com, January 17, 2020, www.washingtonpost.com/religion/2020/01/17/famed-buddhist-nun-pema-chodron-retires-cites-handling-sexual-misconduct-charges-against-group-leader

Dalia Mogahed. "American Muslim women don't need you to save them from Islam. They need your respect," USAToday.com, www.usatoday.com/story/opinion/2018/08/10/racism-islamophobia-hurt-muslim-women-islam-does-not-column/881492002

Own Your Education

"Not everything that is faced can be changed. But nothing can be changed until it's faced."

—**James Baldwin**

***

The presence of difference often sparks a range of emotions in the workplace. When it comes to the difference of a Black person's hair texture and style compared to that of a White person, I find that emotions range from curiosity to outrage. The natural hairstyles of Black Americans continue to be controversial and remain on the front line in the fight for racial equality as hairstyles are often deemed a distraction and inappropriate in the workplace by White people. By contrast, the natural hair styles and texture of White people never seem to so much as raise an eyebrow. For decades, professionalism has been dictated by how well one adapts to European physical features and mannerisms, and people who fall outside that construct must alter their appearance to be regarded as professional.

The intolerance for Black hairstyles serves as a barrier to employment, career advancement, and education, forcing Black people to conform to

53

Eurocentric hairstyles. This requires the straightening of our coarse hair for the sake of economic independence. The Good Hair survey conducted by the Perception Institute revealed that "on average, white women show explicit bias toward black women's textured hair. They rate it as less beautiful, less sexy/attractive, and less professional than smooth hair. [Further], black women perceive a level of social stigma against textured hair, and this perception is substantiated by white women's devaluation of natural hairstyles." Before Black people were enslaved in the United States, the hairstyles of many African tribes symbolized social and marital status, family background, and spirituality. Hair is considered to be an elevated part of one's body. Today, Black people are subjected to uninformed comments, sarcastic questions, and intrusive behaviors concerning their hair and must navigate the best way to respond without drawing further attention while demanding respect and setting personal boundaries.

Dina Neal, the first African American woman elected to the Nevada Assembly, shares her experience in a post with the *Nevada Current*. She explains that when she came to the state capitol in 2011, she wore her hair in braids and that her hairstyle was a practical choice. Carson City was snowy and cold, and she did not want to wear her hair wet while exposed in 25-degree weather. In 2013, a White staffer told her that her hairstyle was unprofessional and inappropriate. The comment stung, but Neal kept her braids. "I just dealt with it. I did not take the four hours to take out my braids," she said in the interview. Eight years after that incident, Neal, now a state senator, saw an opportunity to change the status quo and introduced a bill to prohibit racial hair discrimination. During testimony at a Nevada legislative hearing, Wendy Greene, a professor at Drexel Kline School of Law and a leading authority on legal issues involving race and appearance, stated, "African descendant women and girls [are] in a precarious Catch-22: Either don your natural hair at the risk of lawfully being deprived of employment or an educational opportunity, or don straight hair at the risk of enduring consequential harm to your physical, psychological, economic and physical well-being."

A post from *JSTOR Daily* details the story of Chastity Jones, an African American woman who wore her hair in locs. In 2010 Jones eagerly accepted a job offer from Catastrophe Management Solutions as a customer service representative. The offer, however, came with one stipulation—she had to cut off her locs. When she refused, the company rescinded the job offer. The company's hiring manager reportedly told Jones, "They tend to get messy." The Equal Employment Opportunity Commission (EEOC) filed a suit on Jones's behalf in 2013 and lost. In 2016, the Eleventh Circuit Court of Appeals upheld the district court's ruling and dismissed the case.

Most of us in the workplace don't make hiring decisions or write organizational policy, but we can work to improve the company climate and experiences of Black colleagues who wear their hair in its natural state. When someone changes their hair from short to long overnight via wigs or extensions or wears locs, braids, twists, or other cultural hairstyles, it's natural to be curious. You may find yourself perplexed or fascinated by the shape and texture. Satisfying that curiosity does not mean that it is appropriate to touch your co-worker's hair, ask questions like "What do you call those" and "How long does that take?" Instead, educate yourself. Do the research on how hairstyles reflect culture and the requirement for Black people to assimilate into mainstream to achieve a basic level of acceptance. Asking to touch a Black person's hair infringes on personal space, conjures feelings of otherness, and makes an individual feel like they are on display or viewed as entertainment for their naturally straight-haired counterparts. Black hair has a long history of being politicized and stigmatized in the workplace. For Black people, it's not just about hair. It's about choice and being empowered to be who we are culturally in the workplace. Champions of inclusion recognize their cultural biases, acknowledge that perceptions of professionalism are relative, and choose to respect the whole individual without judgment.

See Figures 10.1 and 10.2.

Figure 10.1: Natural hair

Figure 10.2: Natural hair

Actions

Embrace That Natural Is Natural

Natural is defined as existing or caused by nature. The state of natural hair for one race should not demean the natural hair of another. Expand your view of beauty and professionalism. Binge watch a few movies and documentaries for a 360-degree perspective on the struggles, traumas, and triumphs. Check out:

- *Back to Natural*. View the trailer at https://backtonaturaldoc.com.

- *My Nappy Roots: A Journey Through Black Hair-itage* (2006) directed by Jay Bluemke / Regina Kimbell.

- *Nappily Ever After* (2018). View the trailer at www.youtube.com/watch?v=e-pMVkg7wHg.

You can also add this list to your library: www.blackhairsyllabus.com/filmsdocs.

Reflect When Your Personal Space Was Violated

Everyone has experienced a violation of their personal space. Perhaps someone stood too close in an elevator, literally breathed down the back of your neck while in line, or stepped too close in conversation leaving no room to back up. Julius Fast, author of *Body Language*, explains that a person who violates personal space sends the signal, "You are a non-person, and therefore I can move in on you. You do not matter." Personal space allows us to maintain our perceptions of safety and to protect ourselves. Consider a time when your personal space was violated and reflect on how it made you feel. Did you feel anger? Withdraw inwardly? Aggressively respond? Imagine that happening again and again by people you see every day. Examine what you are feeling in this moment to gain a new perspective.

Create Your Black Hair Experience

As you prepare for your next workday, think about whether your hair in its natural state would be of concern or offensive to co-workers, customers, or clients. Imagine if it were deemed unacceptable or unprofessional. Contemplate how their reactions may impact your day as well as career. Consider comments and reactions you've made or heard on the topic. Decide whether you will react differently in the future and in what way.

Action Accelerators

- **YouTube.com**: *No. You Cannot Touch My Hair!* Mena Fombo, TEDxBristol: www.youtube.com/watch?v=OLQzz75yE5A&t=283s

- **YouTube.com**: *The Psychology of Black Hair,* Johanna Lukate, TEDxCambridgeUniversity: www.youtube.com/watch?v=-yJ17ys m5DY&t=490s

- **Book**: *Hair Story: Untangling the Roots of Black Hair in America* by Ayana Byrd and Lori Tharps

- **Huffpost.com**: "4 Questions About Hair That Black Girls Are Tired of Answering," by Jolie A. Doggett: www.huffpost.com/entry/ black-hair-annoying-questions_1_5c5b3d71e4b08710475a3daf

- **BBC.com**: "How does black hair reflect black history?" by Rumeana Jahangir: www.bbc.com/news/uk-england-merseyside-31438273

Sources Cited

Perception.org. "The 'Good Hair' Study Results, August 2016, https://perception.org/goodhair/results

Lebo Matshego. "A History of African Women's Hairstyles Lifestyle," Africa.com, January 25, 2020, www.africa.com/ history-african-womens-hairstyles

Marcia Mercer. "Banning Hair Discrimination Emerges as Racial Justice Issue," NevadaCurrent.com, www.nevadacurrent.com/ 2021/12/01/banning-hair-discrimination-emerges- as-racial-justice-issue

Chanté Griffin. "How Natural Black Hair at Work Became a Civil Rights Issue," Daily.JSTOR.org, July 3, 2019, https://daily .jstor.org/how-natural-black-hair-at-work- became-a-civil-rights-issue

Bakari Akil II Ph.D. "Personal Space: How Violating Unspoken Rules Can Lead to Chaos - at Least for You," Psychology Today.com, April 11, 2010, www.psychologytoday.com/us/blog/ communication-central/201004/personal-space-how- violating-unspoken-rules-can-lead-chaos-least

Focus on Ability

"Abled does not mean enabled. Disabled does not mean less abled."

—**Khang Kijarro Nguyen**

***

Company leaders are making progress in creating welcoming and positive work environments. As organizations struggle to find talent in many fields, individuals with disabilities are increasingly being recognized as a source of engaged, motivated employees. This focus is long overdue. According to the Bureau of Labor Statistics, 19.1 percent of people with a disability were employed in 2021, up from 17.9 percent in 2020. As we empower people with disabilities to fully participate in a world not designed for them, everyone benefits from their contributions. Like the rest of us, they are unique with a wealth of knowledge, skills, and talents, which adds another dimension of diversity, resourcefulness, and creative energy to the workplace. The Americans with Disabilities Act (ADA) defines an individual with a disability as one who has a physical or mental impairment that substantially limits one or more major life activities. The fact that limitations exist, does not make them any less capable, or prevent them from delivering great work with the use of accessible tools and accommodations. Many have invested most of their lives in creating workarounds so they can comfortably exist and prosper. That vantage point brings fresh perspective and new opportunities. The Society of Human Research Management (SHRM) posted "according to the 2019 Disability Statistics Annual Report from the Institute on Disability, nearly 1 in 8 people in the U.S. has a disability, and that number is rising annually."

When we have co-workers with disabilities, we must change the way we think about them and challenge our beliefs. According to SHRM, more than 66 percent of workers with disabilities have experienced negative bias. Biased attitudes and the fear of saying the wrong thing creates a barrier to relationship building and causes us to feel awkward when communicating. Awkwardness breeds awkwardness. Before you know it, you and your co-worker are feeling awkward. This is a natural phenomenon when we're not accustomed to interacting with someone who experiences visual or hearing loss, uses a wheelchair, or has Down's Syndrome or some other visible disability. Most often, our first reaction is pity. We applaud them for doing the everyday things that most of us take for granted and tell them how inspirational they are for doing so. Saying things like that comes from a good place; however, it is patronizing. The constant drip of patronizing comments can be devastating. The most important thing to remember when engaging is that they are people, not their disability. The disability is just one of their many characteristics. The more we see the individual, the more we'll realize their capabilities.

Make it a goal to understand the concept of disabilities, as recognized by the ADA. It's helpful to not only be aware of the accommodations set by our organizations but also our role in helping disabled co-workers achieve a sense of belonging. This will take conscious effort as we unlearn and interrupt default patterns. We should be aware of everyday language and avoid using disability metaphors like "lame excuse" or "turning a blind eye." Learning to adapt to each situation is how we work toward being more inclusive. For instance, you may have to offer a visually impaired person your arm when walking together or let them know when you are leaving the room. Allow them to decide that they need your help. Don't assume. Respect people with disabilities just as you would anyone else. It's the right thing to do. Focus on abilities—what a person can do rather than what a person cannot do. Champions of inclusion acknowledge, understand, and embrace the widespread nature of different abilities while exhibiting comfort over awkwardness when interacting. Attributing value to the lives of others is a gift we can give every day.

Actions

Acquire Disability Etiquette

Etiquette varies by environment and circumstance and is one of the ways we can demonstrate respect for other people. To acquire disability etiquette, check out the following:

- **CerebralPalsy.org**: Disability Etiquette
- **Apa.org**: Choosing Words for Talking About Disability
- **Vantagemobility.com**: Disability Etiquette: How to Respect People with Disabilities

Create a Lived Experience

Increase your understanding of people living with disabilities. On your next outing or commute to your place of work, take the handicapped-accessible route—no stairs, no doors that don't automatically open, find street curbs with ramps, etc. Notice the time and energy it takes to get to your destination.

Become an Advocate

Align yourself in support of individuals with disabilities by advocating for a disability-focused affinity group or join an existing one as an ally. Volunteer at an organization that supports individuals with disabilities. Help others acquire disability etiquette.

Welcome Feedback

Be open to in-the-moment advice or coaching from colleagues with disabilities when unintentional slights happen. Ask that they alert you right away as you build your disability etiquette.

Action Accelerators

- **YouTube.com:** *If I Can...*, Chris Koch, TEDxBinghamtonUniversity: www.youtube.com/watch?v=ftgxQYPWjX8&t=871s

- **Medium.com:** "I Don't Want to Hear Your Ableist Slurs Anymore," by Lia Seth: medium.com/@LiaSeth/i-dont-want-to-hear-your-ableist-slurs-anymore-229538b9fc80

- **JenniferBrownSpeaks.com:** "C-Suite 2.0: The Chief 'Heart' Officer Role and How Empathy Will Fuel Innovative Cultures of Belonging," by Jennifer Brown: https://stitcher.com/show/the-will-to-change-uncovering-true-stories-of-diversity-and/episode/e38-c-suite-2-0-the-chief-heart-officer-role-and-how-empathy-will-fuel-innovative-cultures-of-belonging-56032184

- **YouTube.com:** *How to build inclusion into the everyday,* Annamarie Jamieson, TEDxAuckland: www.youtube.com/watch?v=ltHMyyfCDqg&t=604s

Sources Cited

SHRM.org. Attracting and Retaining Workers with Disabilities try this one. www.shrm.org/resourcesandtools/tools-and-samples/hr-forms/pages/attracting-and-retaining-workers-with-disabilities-.aspx

Strive to See the Whole Person

"[O]ur attitudes towards things like race or gender operate on two levels. First of all, we have our conscious attitudes. This is what we choose to believe. These are our stated values, which we use to direct our behavior deliberately . . . But the IAT [Implicit Association Test] measures something else. It measures our second level of attitude, our racial attitude on an unconscious level—the immediate, automatic associations that tumble out before we've even had time to think. We don't deliberately choose our unconscious attitudes. And . . . we may not even be aware of them. The giant computer that is our unconscious silently crunches all the data it can from the experiences we've had, the people we've met, the lessons we've learned, the books we've read, the movies we've seen, and so on, and it forms an opinion."

—Malcolm Gladwell

***

Many of us consider ourselves to be well-intentioned, egalitarian, fair-minded individuals who would never allow our biases to negatively inform our interactions with people we work with. Consciously, we are absolutely determined to be kind, objective, and nonjudgmental.

Unconsciously, we often do just the opposite. Meet unconscious or implicit biases. They are triggered involuntarily and without our awareness. The truth of the matter is that while we mean well, the workplace is rife with implicit bias. We pack them up and carry them around with us every day. We can size people up in a matter of seconds and allow those perceptions to affect the way we work and communicate with one another. According to scientists, there is no way around it. Implicit bias is ingrained, and everyone has it. When Karen, a White sales manager enters the office kitchen, she encounters Robert, a young Black man dressed in a hoodie and sneakers, and she instantly concludes that he's part of the cleaning staff. As she sparks a conversation about the weather, she alerts him to the fact that there are no coffee mugs in the cabinet. Perplexed, he asks why he needs to know about the lack of coffee mugs. He explains that he is on the software development team and that he's on his way to a meeting and apologizes that he can't help her with the coffee mug problem. Surprised at the discovery, and slightly embarrassed, she apologizes for the faux pas. In this instance, she didn't intend to judge; she sparked a friendly conversation about the weather and let her bias guide the interaction. Perhaps she has never met a Black software engineer, and she associates hoodies and sneakers with cleaning staff when she sees that attire in the office. The effects aren't always negative. On the other hand, given the same situation, what if she held a point of view that Black men wearing hoodies and sneakers were professionals? After all, Meta's (formerly known as Facebook) Mark Zuckerberg frequently dons hoodies and sneakers—even to his company's initial public offering. Implicit bias could mean that she is more open to the possibility that the man in the kitchen is not only a professional but a high-ranking executive. At least in this scenario, Robert wasn't demeaned, insulted, or diminished as she had clearly not intended. The key reason our biases are so pervasive is that our brains have a natural tendency to seek patterns and associations in an effort to make sense of a complex world. Our backgrounds, previous experiences, and cultural contexts directly impact how we judge, evaluate, and relate in the workplace. These biases can divide us and compromise our ability to establish healthy and productive relationships with diverse groups. Had Karen taken a moment to get to know Robert, the mistake would have been avoided, and Robert would not have left the encounter feeling slighted. The mental shortcuts that we use to assess people and situations allow us to navigate through the day more efficiently while simultaneously preventing us from seeing the whole person or situation. Our snap judgments invariably help some and harm others.

The Perception Institute defines implicit bias in these terms, "Thoughts and feelings are implicit if we are unaware of them or mistaken about their nature. We have a bias when, rather than being neutral, we have a preference for (or aversion to) a person or group of people. Thus, we use the term 'implicit bias' to describe when we have attitudes towards people or associate stereotypes with them without our conscious knowledge." The good news is that we don't have to be victims of our biases as they can be managed once we realize what they are. We make the unconscious conscious and adapt our behaviors.

In the workplace, negative biases impact groups based on gender, race, age, ability, LGBTQ status, and culture but can also manifest when we have preferences to things like Anglo-sounding names or physical attributes such as weight, height, and beauty. When men continuously interrupt or ignore contributions of women, that signals gender bias and the belief that men are superior thinkers. If the work of a Hispanic colleague is constantly overscrutinized by a White colleague, it's indicative of racial bias and the belief without realizing it that Hispanic people are less competent. If you notice that the entire sales team has the physical attractiveness of supermodels, the hiring manager is biased toward beauty. Uncover your personal biases with the implicit associations test. It was developed by scientists, grounded in how our brains process information, and designed to uncover unconscious attitudes and beliefs. Once we become aware of our biases toward various groups, we can begin the work of interrupting thought patterns, making gut decisions, and being more mindful in our communications. Champions of inclusion are in touch with their unconscious biases and stereotypical beliefs and we use this knowledge to ensure that they don't show up during interactions and decision making. In the process, we start to actually see individuals as they are and empower ourselves to cut through biased beliefs. Grab a teammate to join you on this journey—spread the wealth.

Actions

Meet Your Biases

Take the implicit association test at implicit.harvard.edu/implicit/takeatest.html. Once you are more in tune with your biases, write down three ways in which you will interrupt them when they crop up. Be specific.

Stay Current on Inclusion Topics

Subscribe to various podcasts on the topic of inclusion and listen regularly. Consider:

- *Choose Inclusion*
- *Outside the Boxes*, Dr. Pragya Agarwa
- *Women's Hour Daily*, BBC Radio

Action Accelerators

- **Youtube.com**: *We all have implicit biases. So what can we do about it?*, Dushaw Hockett, TEDxMidAtlanticSalon: www.youtube.com/watch?v=kKHSJHkPeLY
- **Book**: *Whistling Vivaldi: How Stereotypes Affect Us and What We Can Do* by Claude M. Steele
- **ChicagoTribune.com**: "'Implicit bias': The problem and how to interrupt it. Plus, the beads test," by Michelle Sharpe Silverthorne: www.chicagotribune.com/opinion/commentary/ct-perspec-black-white-families-implicit-bias-0611-story.html

Sources Cited

The Perception Institute, "Implicit Bias," perception.org/research/implicit-bias

CNBC.com: "Here's how Brunello Cucinelli (and his $5,000 blazers) became Silicon Valley billionaires' favorite designer," www.cnbc.com/2019/06/14/how-bruno-cucinelli-became-silicon-valleys-favorite-designer.html

Aspire to Be an Ally

"And if we are lucky enough to be in a position of power, if our voice and our actions can mobilize change, don't we have a special obligation? Being an ally can't just be about nodding when someone says something we agree with—important as that is. It must also be about action. It's our job to stand up for those who are not at the table when life-altering decisions are made. Not just those people who look like us. Not just those who need what we need. Not just those who have gained an audience with us. Our duty is to improve the human condition—in every way we can, for everyone who needs it."

—**Kamala Harris**

***

Allies in the workplace are essential to inclusive cultures, creating positive experiences for others and advancing one's career. Getting ahead, being valued, or respected without gender, race, or other privileges can seem insurmountable. So, how do you know whether you are an ally? I've seen many explanations, and I find this one to be very thorough. Chikere Igbokwe, founder and diversity, equity, and inclusion consultant at Inclucive, defines an ally as "someone who uses their power and privilege to advocate for others. Allies are not members of marginalized groups. Allies support Black, Brown, and minority employees, colleagues, and friends by understanding what they go through. Allies educate themselves on issues that affect Black, Brown, and minority people, speaking out on injustices, educating colleagues, friends, and family about being an ally and acknowledging their privilege and being actively anti-racist." Taking this a step further, anyone with some level of privilege can be an ally to others with less privilege. As a Black, able-bodied, cisgendered female, I am definitely in a marginalized group. My privilege comes from being able-bodied and never having to give a second thought to my identity. I educate myself on the issues facing individuals with disabilities and my LGBTQ+ friends and work to make a difference.

Everyone has at one time or another needed the support of an ally. Prisha was a mid-level finance manager who wanted to be on the strategic task force for a high-profile project. She had no relationship with the decision-makers, when someone made an introduction that positioned her to articulate her value and ultimately landed her on the team. Hakeem introduced an idea weeks ago that didn't get any traction only to be presented later by a peer who was applauded for it. When Hakeem spoke up to reclaim his idea, someone stepped in to back him. These acts of allyship made the difference between getting what was needed or desired and being left out in the cold with feelings of anger or frustration. Allies have the ability to bring about change for underrepresented groups by using their positions of power and privilege to interrupt the systemic and institutional dynamics that perpetuate inequities.

Allies are not self-proclaimed. We're an ally when the groups we support say that we are. We must build a demonstrated track record of impactful deeds. Being an ally requires courage as we actively engage with the work that drives change, which not only takes us out of our comfort zone but makes others uncomfortable as well. Let's face it, disrupting the status quo from which many have benefited can cause chaos and conflict. Understand that allyship is a journey of commitment for the long haul. The good news is that consistent acts of allyship change lives. As we create a groundswell of people living their best lives, the more we all benefit from the contributions they make.

The role of White men as allies can't be emphasized enough as they are the group with the most privilege and power. A post on Fortune .com reports that "white men account for 72 percent of corporate leadership at 16 of the Fortune 500 companies." The White Men's Leadership study revealed that "Globally, 32 million white men hold leadership position with 6 million in the United States." After decades of increased focus on diversity, equity, and inclusion, the ever-present underrepresentation of women and racial and ethnic minorities in leadership roles demands improved strategies. The Corporate Governance Research Initiative by Stanford Graduate School of Business shows only three Black people, five Hispanic people, and three people who identify as "other" in the Fortune 100 C-suite. Women account for only seven of those positions.

Begin your allyship journey today. Identify your privileges and take note of those around you with less privilege. In addition, take note of who has equal or greater privilege than you and join forces. Use your authority, power, and influence to dismantle barriers and empower others to thrive. It is okay to start small and work your way to bigger and better. Begin by educating yourself on the issues and take stock of your skills, competencies, and relationships that can be leveraged to increase effectiveness. Like any other worthwhile endeavor, allyship requires skill. The more we do it, the better we will become. Don't be discouraged by mistakes. Mistakes are part of the job. You can take the first step in any number of ways. Perhaps you can educate someone from a marginalized group on the politics of promotion within the organization, decline a coveted assignment and recommend a high potential person from an underrepresented group, or speak up when someone shares racist jokes. Champions of inclusion are active allies and keep an eye on the terrain for opportunities to bridge the gap between what is and what should be.

Actions

Educate Yourself on Allyship

Increase your understanding of what it means to be an ally and how allies can make a difference. Review the characteristics of allies in Figure 13.1. Consider two co-workers whom you believe could benefit from an ally. Observe how current dynamics are preventing them from thriving. Figure out ways you can impact those two individuals. Your efforts will impact not only them but others just like them. No need to announce it. Just do it.

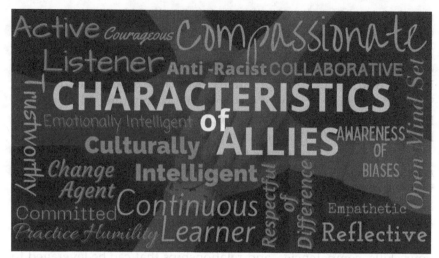

Figure 13.1: Characteristics of allies

Become an Advocate for Organizational Change

Review your organization's employee handbook and look for opportunities where you can advocate for change. Are the biases of leaders creeping into organizational policies and practices? Identify one area where you could potentially make an impact and start a conversation with the appropriate people to open minds toward change.

Create a Network of Aspiring Allies

Spread the wealth. Introduce/connect people from varying privilege. Let them know why you think they should meet. Where people have commonalities are the best opportunities.

Support Others in Living Company Diversity Values

Note organizational diversity statements and commitments. Be mindful of how colleagues and leaders live these values. Support them in their inclusion journey. Practice "calling in" if needed. See the Inc.com activity accelerator in "Action Accelerators" to learn more about calling in.

Make an Ongoing Commitment

If your company has affinity groups in support of traditionally underrepresented groups, join one as an ally. Affinity groups meet regularly and welcome allies in support of their mission, goals, and values.

Action Accelerators

- **CamilleStyles.com**: "How to Be an Ally—3 Questions to Turn Allyship into Action": https://camillestyles.com/wellness/3-questions-to-turn-allyship-into-action

- **SHRM.org**: "Workplace Allies Serve as Ambassadors for Change": www.shrm.org/resourcesandtools/hr-topics/behavioral-competencies/global-and-cultural-effectiveness/pages/workplace-allies-serve-as-ambassadors-for-change.aspx

- **JenniferBrownSpeaks.com**: *Good Guys: Beyond Intent to Impact with David Smith and Brad Johnson,* Jennifer Brown: https://stitcher.com/show/the-will-to-change-uncovering-true-stories-of-diversity-and/episode/e143-good-guys-beyond-intent-to-impact-with-david-smith-and-brad-johnson-80516125

- **YouTube.com**: *Screaming in the Silence: How to be an ally, not a savior,* Graciela Mohamedi, TEDxBeaconStreet: www.youtube.com/watch?v=d2qAbp-t_FY&t=333s

- **Inc.com**: "Calling in Vs. Calling Out: How to Talk About Inclusion," by Maya Hu-Chan: www.inc.com/maya-hu-chan/calling-in-vs-calling-out-how-to-talk-about-inclusion.html

Sources Cited

Alexandra Marinaki. "Allyship in the workplace: Be color brave, not color blind," Resources.workable.com, resources.workable.com/stories-and-insights/allyship-in-the-workplace

Stacy Jones. "White Men Account for 72% of Corporate Leadership at 16 of the Fortune 500 Companies," Fortune.com, fortune.com/2017/06/09/white-men-senior-executives-fortune-500-companies-diversity-data

WhiteMensLeadershipStudy.com. Executive Summary, page 4, www.whitemensleadershipstudy.com

GSB.stanford.edu. "Fortune 100 C-Suite Organizational Charts," February 2020, www.gsb.stanford.edu/sites/gsb/files/fortune-100-c-suite-organizational-charts-feb-2020.pdf

Create an Inclusive Experience

"Diversity is being invited to the party. Inclusion is being asked to dance."

—**Verna Myers**

***

Everybody looks forward to company events. There's the excitement of experiencing a nice venue, hanging out with co-workers in a relaxed atmosphere, making new acquaintances, and being enticed by the food spread. They are an opportunity for leadership to show appreciation for the hard work of staff, enhance company culture, and build relationships. So, why not bring people together and connect for a "Wine Down Wednesday," or "Yoga in the Park" or celebrate a new product launch or the holiday season? The intent is for everyone to have a good time.

In practice, however, not everyone gets to have an enjoyable experience. The thought of attending a company event may secretly spark dread for some employees. And, no wonder, often events are not designed with them in mind. Diverse groups have diverse needs. Organizations that host the same type of events year after year without regard to changing employee demographics and lifestyles are missing an opportunity to demonstrate that they value diversity. Planners must be mindful and work to ensure that gatherings are respectful, welcoming, and open to everybody.

Anna has been an account manager at a midsize tech company for almost a year. She recalls the excitement of her first day and how friendly her teammates were. The open office environment and natural light energized her. Within her first month, the quarterly sales meeting rolled around. Everyone flew into the home office for two days of product updates, training, and strategy sessions. Naturally, there was a breakfast spread. On the first day there was dairy-based yogurt, bagels, and muffins. Anna had to skip breakfast as she is vegan with a sensitivity to gluten. As the morning progressed and hunger intensified, Anna found it hard to concentrate and eagerly awaited lunch. When the catered lunch was served, she was first in line. It was a beautiful array of tri-color pasta, roasted chicken, Caesar salad, French onion soup, and brownies for dessert. Nothing that aligned with her diet. To take the edge off, she filled her plate with the salad, removed the croutons, and scraped off as much of the parmesan cheese as she could. At the table, her teammates couldn't help but notice how painstaking it was to prep the salad before she took the first bite. When she explained, one colleague commented, "Oh, you're one of those." Another quipped, "I don't know how you survive on rabbit food." Embarrassed, she laughed along. Toward the end of the day, she inquired about the following day's menu. The office manager informed her that it was going to be pretty much the same, and it was too late to adjust the order. The next morning, Anna was prepared, at least for breakfast. She picked up fresh fruit from the building cafeteria and ate it before the start of the meeting. The lunch menu hadn't changed much—glutenous choices, meat protein, and not a plant-based item in sight that wasn't covered with something with gluten or dairy. Refusing to torture herself for the second time and perform surgery on her meal, she decided to have lunch in the building cafeteria. When the meeting ended on the last day, Anna asked the admin who orders the food if she could please add vegan choices for the next meeting. The admin gave her word. Imagine Anna's disappointment when the next meeting rolled around, and breakfast was exactly the same as before. She pondered whether she should say something but decided it best to let it go. When lunch was served, she dodged the catered meal and opted for lunch in the cafeteria. She chose to eat alone, rather than return to the group table with her "special" food. She didn't want the attention. With the exception of her private lunches, Anna was fully present at all sessions quarter after quarter. When it was time for her performance review, Anna expected that it would go well. She exceeded quota, and clients really liked her. As she and her manager discussed her performance, he gave her a low score as a team player, citing that she was aloof, particularly at team meetings and advised her to work at connecting better with colleagues. Further, he encouraged her to take advantage of the fun atmosphere at team lunches. She was appalled at his comments and dismayed that her vegan lifestyle would affect her performance review. It made her feel like an outsider because of dietary restrictions, and the snide, insensitive remarks she endured from colleagues were cause for concern.

It's not unusual that organization meeting and event planners don't have the knowledge and information they need to create an inclusive experience for all. Such responsibilities tend to be assigned to admins who do this work in addition to the dozens of other projects they juggle. Inclusive events don't happen by accident, rather, by intention. Involving other staff of diverse backgrounds and lifestyles can be helpful, but only if there is a stated goal to be inclusive. Otherwise, there are just more hands on deck perpetuating exclusion. The added perspective can generate new ideas, identify opportunities to be more inclusive, and ensure maximum attendance. When organizing, bake the following ingredients into the discussion:

- Time of day that allows staff with small children or aging parents to attend
- Avoid dates that conflict with religious holidays
- Accommodations for those with disabilities
- Nonalcoholic beverage options
- Gender-neutral restrooms
- Dietary preferences such as kosher, gluten free, or vegan
- Inclusive language in the invitation messaging
- Use of diverse caterers, vendors, and minority-owned venues
- Follow-up survey for continuous improvement

Although most of us are not involved in the planning of company gatherings, that does not mean that we don't play a role in the overall experience of attendees. Little things can make a big difference. Make it a point to interact meaningfully with co-workers who are different from you. Seek common identities like parents, sports or art lovers, or travelers to fun or exotic places and expand from there. Listen more than you talk. Offer to introduce folks to others, be sure to say names correctly while being mindful of gender pronouns when doing so. Be aware that humor is not the same across cultures. Something we may find funny and harmless may be insulting to someone else. When chatting with co-workers who talk with an accent, speak with clarity and monitor your tone. Many of us will default to speaking louder rather than more succinctly. Strive to ensure shared understanding. And, as in Anna's story above, please don't make people feel uncomfortable enjoying their meal. Respect food choices. Everyone should be able to enjoy the same quality of experience. Champions of inclusion develop a cultural lens from which to interact and avoid making assumptions based on our own world view. We ask questions, listen for the response while checking and rechecking for bias creep. As we realize the effort toward creating an inclusive atmosphere, the more open we are to requesting accommodations to ensure a positive experience for all. It's time to open the door to more inclusive events.

Actions

Recognize Missed Opportunities

Reflect on your last company-wide or all-hands team gathering. Note who was fully engaged, who was on the sidelines, or who left unusually early. Consider the potential causes that people did not engage. If you have a relationship with those individuals, ask them why. If there was a missed opportunity to be inclusive, talk to company organizers.

Participate in the Planning Process

Join or create an event planning committee representative of diverse backgrounds and experiences. Involve the rest of the organization by creating a suggestion box. This brings remote workers into the process.

Stop Chaos Before It Starts

Ensure or advocate for a "code of conduct" policy so that all have a positive experience. If the event code of conduct has not been updated recently, now may be a good time to review it.

Strengthen the Connection

After a gathering, we tend to return to business as usual and don't keep in touch with new acquaintances. Champions of inclusion keep the connections going in fun ways. Consider sending articles or memes reflective of shared interests and hobbies, creating a memory board of photos and selfies taken and post to a slack channel for continued conversation long after the event ends.

Action Accelerators

- **JournalofAccountancy.com**: "How to Make Company Events More Inclusive," by Cheryl Meyer: www.journalofaccountancy .com/newsletters/2016/may/make-company-events-more-inclusive.html

- **TheInclusionSolution.me**: "Let's Get Practical: A Checklist for Inclusive Meetings and Events," by Leigh Morrison: www.theinclusionsolution.me/lets-get-practical-a-checklist-for-inclusive-meetings-and-events-diversity-and-inclusion-strategy

- **TeamBonding.com**: "12 Reasons Why Team Building Works," by Amanda Deiratani: www.teambonding.com/6-reasons-for-team-building

- **HBR.org**: "How Managers Can Make Casual Networking Events More Inclusive," by Ruchika Tulshyan: hbr.org/2018/10/how-managers-can-make-casual-networking-events-more-inclusive

Disrupt Power

"Inclusivity means not "just we're allowed to be there," but we are valued. I've always said: smart teams will do amazing things, but truly diverse teams will do impossible things."

—**Claudia Brind-Woody**

***

Every organization, every team has a power structure. When we examine the organizational chart, the names and titles in the boxes have authoritative power based on position. Then there are those who wield informal power. They tend to be well-connected and have unearned privilege, experience, or a dominant personality giving them influence with others by the stature of their position in society. Established norms and company culture positions them to lead and be respected. Whether formal or informal, power systems manage and control employee behavior through intimidation or inspiration. Power is primarily shaped by diversity and equity dynamics. A quick check of company websites and annual reports paint a clear picture of who is in control. Images of the C-suite, senior executive team, and board members of companies in Western cultures are consistently of the same pedigree: white males older than

40, assumed heterosexual, and from middle- or upper-class socioeconomic backgrounds. With the increased focus on diversity, there may be a spot or two for people of color that offers the illusion of inclusion and power when equity is not embraced.

These power dynamics carry themselves into meetings and are quite apparent when we understand the importance of who sits where. The rules are unwritten but understood by all. Conference rooms hold rectangular tables with rows of chairs on either side, and a single seat on both ends. The person with the most authoritative power sits at the head, the lone seat at the end, closest to or facing the door. This vantage point gives command and control of the room as one can see everything from who is coming in or going out to who is engaged or distracted or who has bought into the message and who questions it. If you are not the most powerful person in the room, taking this seat signals that you are. The seats to the left and right of the head are reserved for allies and the most trusted advisors. The lone seat at the opposite end is usually taken by influencers, those with informal power, and the seats to the left and right of them are intended for their allies. The seats in the middle are for the collaborators. They contribute to and support agenda items and are often tasked with doing the actual work. People who prefer to go unnoticed will find a spot in the middle seats or take seats away from the table. Whether we recognize it or not, we should assume that power dynamics are always present in meetings. Teams gather to create, innovate, or solve problems, and they present an opportunity to raise one's profile. However, this psychological geography bends toward the belief that the best ideas come only from those in the power seats. On the contrary, the best ideas can come from anywhere in the room. How often have you seen an idea emerge from the center of the table but does not get any traction until someone in a power seat repeats it and takes credit for it? Disrupting the power at meetings allows all of us to contribute, feel heard, and be respected and can sometimes be as simple as altering who sits where. Consider ways to mix up seating arrangements. Offer to trade your power seat or invite co-workers sitting on the fringes to come closer so that everyone can hear what they have to say. Let's not forgot about our remote colleagues. With remote workers being here to stay, hybrid meetings are now the norm. Organizations and teams need to introduce and develop new habits that are both inclusive and empower those not in the room to contribute. Without targeted intervention, the power dynamic of on-site teams has the potential to be amplified in remote settings. Best practices for meeting room setups include the following:

- Use a central videoconference screen.
- Ensure Internet connectivity is robust enough to support video as well as audio.
- Use ceiling- or table-mounted microphones that cancel background noise so remote colleagues can hear and be heard.
- Technology should accommodate screen sharing, annotating, whiteboarding, or the equivalent.
- Set rules of engagement.
- Consider time zones when scheduling.
- Assign a point person who monitors the chat.
- Periodically ask for questions and comments.
- Acknowledge contributions and note who is and is not contributing— introverts may be participating via the chat.
- Add an agenda that notes who is covering what so all are prepared to contribute.
- Consider who is not on the meeting invite but should be.

You'll know efforts are working when conversations are flowing, and all are engaged. No one is afraid to speak up. Connections are strengthened, and we get to maximize our potential. FastCompany.com advises in a post on leadership, "A players (leaders who share power) hire people better than themselves and give them support to excel. Then they give them recognition and help them move up. The whole organization gets stronger." It continues, "Real leaders communicate a lot. They make it a point to share as much information as possible with everybody. They see additional power in having a well-informed team that can contribute more because they know more. Effective leaders win people over by building an open environment of trust and respect. They create meaning for people so they can feel proud of their work. They offer personal recognition. They go out of their way to make the work matter to the people doing it."

Champions of inclusion give and share power as well as empower others to participate fully Let's work to level the playing field and start having conversations about power.

Actions

Ensure That Everyone Can Be Heard

Everyone should have the opportunity to be heard. Some colleagues may be apprehensive or too polite to insert themselves into the discussion. Before the meeting, connect with them to let them know that you believe that they can add value and you will be inviting them to contribute. Ask if they are okay with that. Champions of inclusion create safe spaces and leave room to say no when a request makes someone uncomfortable.

Take note of who does not contribute during meetings and have a casual and sincere conversation as to why they don't—no judgment here. There's no right or wrong answer. Listen for cues where you can offer advice or advocate for change in meeting dynamics.

Decrease the Visibility of Power Dynamics

Explore whether there's an option for meeting rooms to be reconfigured to where the power dynamic is less visible.

Encourage Sharing of Power

Educate colleagues that power can be shared for the good of all where everyone benefits. Start a discussion on how to share power and create a plan that all agree to. Make sure it's accessible for all to see.

Action Accelerators

- **Atlassian.com:** "How to counteract 3 types of bias and run inclusive meetings," by Hilary Dubin: www.atlassian.com/blog/teamwork/how-to-run-inclusive-meetings

- **DevelopmentGuild.com:** "How to Lead Inclusive Meetings": www.developmentguild.com/dei/how-to-lead-inclusive-meetings

- **HBR.org:** "Who's Being Left Out on Your Team?" by Carolyn O'Hara: hbr.org/2014/08/whos-being-left-out-on-your-team

- **RallyBright.com:** "Understanding Team Dynamics": www.rallybright.com/understanding-team-dynamics

- **YouTube.com:** *Change how teams work together by eliminating power dynamics,* Advisory Board: www.youtube.com/watch?v=RrEhiwtzhVY
- **InteractionInstitute.org:** "Power Dynamics: The Hidden Element to Effective Meetings": https://interactioninstitute.org/power-dynamics-the-hidden-element-to-effective-meetings

Sources Cited

Patty Azzarello. "Why Sharing Power at Work Is the Very Best Way To Build It," FastCompany.com, www.fastcompany.com/3004867/why-sharing-power-work-very-best-way-build-it

Build Equality into the Day-to-Day

"Inequality, in all its forms—gender, LGBTQ, racial, or otherwise—is an issue that every company must address for its own benefit and to create a better world. We believe businesses need to focus on closing the equality gap with the same energy put into creating new products and markets."

—Tony Prophet

***

Many of us begin our careers with sheer optimism and drive. We were hired because of our knowledge, skills, and expertise and the expectation that those abilities will contribute to organizational goals. There's no reason to believe anything to the contrary (just yet). We enter the workplace believing that we have equal access to opportunity and that working to the best of our ability leads to success. When it comes to women, however, gender inequality can quickly derail career aspirations when equality and equity are not prioritized. When we layer ethnicity into the equation, institutional systems favor their White counterparts in support of career advancement and place historically marginalized groups at a disadvantage.

Many believe that equity and equality are interchangeable when talking about diversity. Consider equity the building block of equality. When we think of equality, we think fairness. What we do for one person or group is exactly what we do for another. In the workplace, you may have experienced a new mother having six weeks of maternity leave, while new fathers don't get any. As company policies adjust to include new fathers, the policy becomes fair. New mothers and new fathers are getting the exact same thing. Equity is recognizing that we all don't need the same thing. The unique needs of the individual are considered in an effort to acknowledge and correct imbalance so that they too can thrive. Equity in the workplace may include a policy that provides hybrid, job sharing, or flexible hours to accommodate employees who care for small children or aging parents. The ability to get work done anytime, anywhere positions them to continue to achieve company goals and objectives as well as fulfill family responsibilities. When we put equity first, equality follows.

It's not unusual to find that women and, in particular, women from underrepresented groups are assigned the menial tasks that keep the team humming, including capturing meeting minutes, distributing them, scheduling the follow-up, ordering lunch, and other busy work that not only does not support career advancement but also takes time away from activities that do—unless those specific duties were what they were hired to do in the first place. Couple this with being ignored or constantly interrupted during one-to-one discussions and group settings. Chronic disruptions to high value work and consistent interruptions in conversation wreak havoc on relationships and productivity.

Studies have shown that men interrupt 33 percent more often when speaking to women than when they speak to other men. Researchers found that men interrupted women 2.1 times over the course of a three-minute conversation and, during conversations of the same duration, interrupted other men only 1.8 times, while women on average interrupted men only once. The *New York Times* reports that virtual meetings can mirror the same inequities of in-person meetings for women and are starting to crystallize how much harder it is for them to be heard. These inequalities are often not intentional but rather based on gender stereotypes, historical and socialized gender norms where men have the dominant position, and the difference in upbringing of sons versus daughters.

Girls tend to be nurtured from an early age on the importance of being liked, kind, and polite, and carry that mindset into the workplace—a mindset of values that carries little weight in career advancement decisions; while boys are reared to be assertive, competitive, and goal-oriented without much regard for being liked. Growing up, their value was derived from being effective at winning. For them, being the first one picked to join the sports team meant they were trusted to win, while being the last one picked was embarrassing and signaled being a loser. In elementary school, the girls always heard, "It's not whether you win or lose, it's how you play the game." By contrast, the boys were pushed to win, to stretch themselves, and to learn and grow from it. It built their confidence and competitive spirit. For them, doing anything "like a girl" was an insult and implied being ineffective. Media portrayals of women and men reinforce this line of thinking and affect our behaviors and attitudes in the workplace. It's no wonder that gender role expectations map into our day-to-day and how we work together.

Many would say that women have taken long strides toward achieving equality in the workplace. Gone are the days where women were believed to not have what it takes to make it in the world without a man. Women no longer go to college for the sole purpose of finding a husband, and some have figured out how to scrape themselves from the sticky floor and break through the glass ceiling. However, the data say it's not nearly enough. A post from WhatToBecome.com shares:

- 14.6 percent of board chairs belong to women.
- 8 percent of the Fortune 500 chief executives are women.
- 26 percent of employees in the computing workforce are women.
- 29 percent of senior management positions in the world are held by women.
- 52 percent of women in the world deal with noninclusive behavior at work.
- 4 percent of women say their organizations have made progress in building inclusive cultures that support women.

Changing these statistics means changing systems, behaviors, and attitudes. Men supporting women and women supporting women. Our success in creating workplaces in which women have equal access to opportunities they need to thrive depends on our willingness to confront the history and impacts of structural racism and learn how implicit bias and explicit bias operate—the unconsciously or consciously held beliefs and attitudes that shape how we evaluate or behave toward members of a particular group. When we are intentional, our day-to-day actions can in effect interrupt inequitable practices at the interpersonal, institutional, and structural level. If you sit at the table where decisions are made, you have the power to make space for deserving women. I have never met a man who did not want for his wife, daughter, sister, mother, or aunt to succeed. Yet, they intentionally or unintentionally behave in ways that hold back that of another and soothe the blow with comments like, "It's not personal, it's just business," as if that gives permission. Women can be both one another's best friend and worst enemy regardless of their position in the organization. The strong bonds fostered between women can be integral to career success and well-being when the other's best interest is top of mind. Conversely, I can't think of anything more destructive than catty behaviors that women exact on other women resulting in toxic work environments. Cattiness is defined as mean-spirited vicious acts intended to harm. I've personally experienced and heard countless stories of torment based on jealously, insecurity, and desire for power. Everyone has the ability to address gender inequities and inequality at varying levels. Review activities one and two to reflect and leverage your why. We can no longer do things the same way, with the same people, and expect equitable behaviors and systems to emerge. Champions of inclusion are intentional to disrupt the norms that favor the dominant group and communicate how unique approaches add value to the team. We'll find that as we implement change on behalf of one group, other groups are positively impacted. Everybody benefits.

Actions

Create an Equitable System for Completing Menial Tasks

Menial tasks are important but should not be assigned to individuals where those responsibilities fall outside of their job role. Take note of who carries out low-value tasks. If it's not part of the individual's job description, collaborate with colleagues to create a more equitable system.

Set an example. Offer to assist the individual who is stuck with office housework. Don't announce it. Just do it. Invite others to assist. Do this consistently until it becomes the norm.

Assess Team Norms for Bias

Established teams have established practices. Ask yourself who benefits from team processes as well as spoken and unspoken rules, who has the likelihood of being harmed, and what can you do to adjust the imbalance and avoid harm in the future.

Address Imbalanced Communications

Everybody wants to be heard and allowed to contribute. Before interrupting, pause to consider the implications: Can your thought/question wait until the speaker has completed her thought? Would taking notes be better to be followed up on later?

Interrupt the interrupters with phrases like, "I would like to hear Nancy's thought before we continue." Create a few phrases to keep in your back pocket.

Elevate and Amplify the Work of Women

Women want fulfilling careers just as men do. Capable and competent women need others to know about the great work they are doing but may lack connections and influence. Take action to:

- Leverage your power and influence to position deserving and high potential women for success.
- Offer your speaking opportunity.
- Recommend for a high-profile project or stretch assignment.
- Make it a common practice to connect and collaborate with remote workers.
- Acknowledge and encourage talent wherever you see it.
- Foster women mentoring men relationships—both will benefit.

Action Accelerators

- **HBR.org**: "Women of Color Get Asked to Do More 'Office Housework.' Here's How They Can Say No (Difficult Conversations)," by Ruchika Tulshyan: hbr.org/2018/04/women-of-color-get-asked-to-do-more-office-housework-heres-how-they-can-say-no
- **Marketwatch.com**: "Already paid less than men, women are still asked to do the 'office housework'," by Karl Paul:

marketwatch.com/story/already-paid-less-than-men-women-are-still-asked-to-do-the-office-housework-2018-10-08

▪ **FastCompany.com**: "How to End the Office Housework Gender Bias," by Lydia Dishman: fastcompany.com/3050043/how-to-end-the-office-housework-gender-bias

▪ **CCF.org**: "Making Gender Equity in the Workplace a Reality": ccl.org/articles/leading-effectively-articles/gender-equity-workplace

▪ **NPR.org**: *What 'likeability' really means in the workplace*, Andee Tagle and Clare Marie Schneider: www.npr.org/2021/06/15/1006695654/women-gender-bias-work-likeability-career-advice

▪ **Ted.com**: "The Myths that Hold Back Women at the Workplace," Star Jones: ted.com/talks/star_jones_the_myths_that_hold_back_women_at_the_workplace

Sources Cited

Advisory.com. "How often are women interrupted by men? Here's what the research says,", October 30, 2018, https://advisory.com/daily-briefing/2017/07/07/men-interrupting-women

Alisha Haridasani Gupta. "It's Not Just You: In Online Meetings, Many Women Can't Get a Word In," Nytimes.com, April 4, 2020, https://nytimes.com/2020/04/14/us/zoom-meetings-gender.html (subscription required)

WhatToBecome.com. "28 Shocking Women in the Workplace Statistics [2021 Data]," August 10, 2021, https://whattobecome.com/blog/women-in-the-workplace-statistics

INTERVIEW: TELLING IT LIKE IT IS. . .NOT AS WE WISH IT TO BE

Sexism Spins the Employee Revolving Door

Meet Valerie. She has just started her career as a technical support specialist at a state agency in Texas. She is on a team of about 50 people, one of only a handful of women and also one of the youngest on her team. She wasn't surprised by the gender imbalance as the field of technology has long been dominated by men and is often characterized as having a bro-like culture—a subculture of cisgender men who spend time socializing with others much like themselves and excluding those who are different. The bro culture of the tech team combined with the old boys' network of the state agency pretty much guaranteed that there would be a very narrow view of women and that sexist and chauvinistic views would guide most interactions between male and female co-workers. A few months into her role, on a Friday afternoon, two male colleagues invited her to lunch. Valerie, in her kind and friendly spirit, responded, "Sure! I'd be happy to join you." It was a simple exchange—she was invited, and she accepted. It turned weird when one guy changes the enthusiasm in his tone by saying, "Oh, it's Friday." Puzzled by the reaction, she inquired, "Yeah, so what it's Friday—why does that matter?" That's when he suddenly remembered that Friday is strip club Friday—a regular every Friday occurrence. In that moment Valerie realized that they never expected her to join them, which was fine by her as there was no way she would ever lunch at a strip club with co-workers. She was shocked and somewhat appalled at the idea that in a professional environment, colleagues would behave in such a manner, that they considered it acceptable, and that perhaps their behavior was even embraced by the rest of the team. The thought of them partying at a strip club for lunch and then returning to work led her to question who else knew and why it was seemingly condoned. The more she thought about it, the less likely she would be to ever join them for lunch under any circumstance. This single interaction helped her make sense of other actions that went against her standards of teamwork, professionalism, and integrity that she had ignored up until this point. She noticed how their camaraderie was often at play when they validated each other's ideas in meetings while not considering and sometimes stealing ideas of others, or when they made excuses for one another's bro behavior and explained it away as co-workers overreacting. As she reflected on this experience, it made her realize the lack of integrity that permeated throughout the organization. She reflected on the fact that when she negotiated a week's vacation in advance of her start date, it was later reneged when a junior colleague complained to the hiring manager that it was against policy and that if one person did not have to abide by the rules, neither should anyone else. The agency's vacation policy required hours to be accrued before being eligible to take the paid time off. Valerie's intention was not to take advantage of the situation as her trip had been planned well in advance of her employment. Rather, she would just not take the vacation once accrued since it would have already been used. However, the hiring manager demanded that she work longer hours to include Saturdays to accelerate the timeline. That remediation came across as unfair and unreasonable and spoke volumes about not only the character of the leader but the integrity of the agency. The situation left Valerie feeling that leadership could not be trusted and wondering what type of senior leader would not keep their word and leverage their authority when challenged by a junior associate for a decision that they were at liberty to make. Now that Valerie was awakened to the environment, she realized that it was not a place where she would stay long-term...and she did not. She resigned a few weeks later. She was gone within five months of being hired. The environment was simply not a fit and did not align with her values of how women should be treated and how leaders should lead.

Devote Yourself to Critical Thinking

"Do not be guided by your dull eyes, nor by your resounding ears, but test all things with the power of your thinking alone"

—**Parmenides**

***

Everyone arrived in this world with the blank slate of innocence. We had no labels or judgments for others, and our thoughts and actions were genuine. There were no preconceived notions or assumptions. We did not decide that we didn't like broccoli just by looking at it; we waited until we tasted it before deciding. When we were curious about something, we asked a thousand questions until the curiosity was satisfied, often to the exasperation of our parents. The kids next door were just the kids next door, and it didn't matter how much we differed as long as we had fun while together. As time passed, we adopted the beliefs and attitudes of our parents. The strength of our relationships with others thrived or died based on their approval. The more we mirrored them—good or bad—the more love and acceptance we received. Our world expanded when we began formalized education. Textbooks told us stories of early civilization and world history as well as taught us economics and government policy.

We formed new opinions and beliefs and accepted them as fact. While we may have debated and challenged the information back then, it still informs our thinking and actions today.

Humanity believed that the earth was flat until approximately 300 B.C., when Eratosthenes, a Greek mathematician, developed a method of calculating the circumference of the earth, which evidenced that the earth was round. He challenged the belief and proved it inaccurate. The Holy Bible describes Jesus as having skin like "burnished bronze" and "hair like wool," yet every image displays the contrary, a man with white skin and long straight hair. Raised a Christian and educated primarily in Catholic schools, I was taught that one can't pick and choose which parts of the Bible to believe. Believe all or none as every word is true, and if one word was false, it rendered the entire book false. Teachers could never provide a satisfactory explanation regarding the contradiction in the biblical description of Jesus and the pictorial representations. My conclusion: imagery matching the biblical description is unacceptable to the dominant culture, so they created images in their likeness, making them acceptable.

Growing up, teachings of the explorer Christopher Columbus hailed his explorations, and we celebrated his discovery of America with a day off from school. As an adult, I learned the rest of Columbus' legacy—of his enslavement, outright theft, and the genocide of indigenous peoples. This part of the story was not told in my textbooks. As more Americans realize the harsh reality of our country's historical racism, Columbus Day remains a federal holiday, and folks happily flock to parades and celebrations in his honor.

Many people believe and act on political rhetoric like the labeling of Mexican immigrants as rapists and criminals, Black people as violent and lazy, and Haitians as having AIDS. BBC.com reported in May 2021 that hate crimes in the Asian community rose by 1200 percent as Asians were targeted and blamed for the covid-19 virus. The long-held and incorrect belief persists that members of the LGBTQ community could heal themselves of homosexuality if they just tried hard enough. The fact of the matter is that there are rapists, criminals, AIDS victims, and those who are lazy and violent in all races, yet only marginalized groups are called out and subsequently victimized for it at a much higher rate than those of the dominant culture. My gay friends and family members will tell anyone who asks that their sexuality is no more a choice than that of heterosexuals. I have yet to meet a heterosexual who made a conscious decision to be so. Distorted views and fallacies presented as fact create false perceptions of diverse groups and cause harm, even death. Developing an inclusive mindset requires us to think differently and devote time to critical thinking. Maintaining ignorance and rigidity in our beliefs and opinions causes us to stay stagnant in our understanding. Critical thinking requires the ability to engage in reflective and independent thinking.

We must gather and synthesize information from multiple sources as an active learner, rather than an uninformed consumer of information. We can choose to examine our beliefs and then choose to change them.

As champions of inclusion, we are on a learning journey. We should constantly assess what we think we know about groups, individuals, and institutional systems, and challenge our assumptions. What was true yesterday, last month, or last year may not be relevant today. Our world is changing so quickly, and we need to consider new variables that arise and that often serve as barriers to inclusion as well as create forces resistant to change. Just recently, as I write this book, the state of Texas outlawed abortion after six weeks of pregnancy, knowing that most women won't discover the pregnancy until around the sixth week. Further, a woman's right to choose was completely abolished shortly thereafter when the U.S. Supreme Court overturned Roe v. Wade after 49 years of enacting it into law. Prior to Roe v. Wade, women subjected themselves to illegal abortions or consumed harmful drugs claiming to induce abortion where many suffered and died as a result. Regardless of what we think about abortion, inclusion means thinking about the impact to female colleagues. The state of Florida introduced a bill banning discussions of sexual orientation and gender identity in schools that in part requires school officials to disclose to parents information confided in them about a student's sexual orientation and gender identity—essentially outing LGBTQ youth against their will with potentially harmful effects. In these instances, critical thinkers are not focused on the laws, rather, their impact, and seek creative ways to make a difference. Controversial changes bring a barrage of opinions, facts, and so-called facts that inform our interactions. We've got to determine whether the information is fact or opinion, whether all facts were provided, and whether anything was left out. Champions of inclusion who employ critical thinking make better allies and can model different behaviors for others to emulate.

While we are well past the days of our childhood innocence, returning to our willingness to examine before judging and to be curious in all things is paramount as we strive toward inclusion.

Actions

Strike a Balance

As agents for change, we tend to focus on addressing problems that create barriers to inclusion and that can lead to burnout. Most of us are wired to solve problems as it is a highly sought-after skillset in the workplace. In addition to addressing the issues, leverage critical thinking skills to uncover opportunities for inclusion.

Collaborate with Other Inclusion Champions

Keeping a pulse on the issues and narratives affecting diversity, equity, and inclusion is time-consuming. When we collaborate with others, we gain more insight through diverse perspectives, resulting in greater impact.

Lead by Example

Use actions and influence to attract more inclusion champions. Model the behaviors that you want to see in others. You don't need to have a leadership title to lead. When you can mobilize others and do incredible things, you are, in fact, leading.

Develop a Framework for Critical Thinking

To be effective in critical thinking requires a structured process. A post from the University of the People recommends the following:

- **Define your question**: When it comes to critical thinking, it's important to always keep your goal in mind. Know what you're trying to achieve, and then figure out how to best get there.
- **Gather reliable information**: Make sure that you're using sources you can trust—biases aside. That's how a real critical thinker operates!
- **Ask the right questions**: We all know the importance of questions, but be sure that you're asking the right questions that are going to get you to your answer.
- **Look short- and long-term**: When coming up with solutions, think about both the short- and long-term consequences. Both of them are significant in the equation.

▪ **Explore all sides**: There is never just one simple answer, and nothing is black or white. Explore all options and think outside of the box before you come to any conclusions.

Action Accelerators

▪ **uoPeople.edu**: "Why Is Critical Thinking Important? A Survival Guide": uopeople.edu/blog/why-is-critical-thinking-important

▪ **YouTube.com**: *Penny for Your Critical Thoughts?*, Darcy Roland, TEDxNewport: youtube.com/watch?v=qdJIwIMQwbg

▪ **Grainger.com**: "Five Strategies to Lead Without Authority": grainger.com/know-how/business-operations/people-management/kh-lead-without-authority-five-strategies

▪ **YouTube.com**: *Leadership Has Nothing to Do With Rank*, Simon Sinek: youtube.com/watch?v=LDQj-sU9Opw

Sources Cited

BBC.com: "Covid 'hate crimes' against Asian Americans on rise," May 21, 2021, https://bbc.com/news/world-us-canada-56218684

Look Beyond Appearances

"We need to give each other the space to grow, to be ourselves, to exercise our diversity. We need to give each other space so that we may both give and receive such beautiful things as ideas, openness, dignity, joy, healing, and inclusion."

—**Max De Pree**

***

Appearances can be deceiving as we tend to make decisions and judgments based on very limited information. Visual observations provide only surface details, often by design. Men may wear caps to cover baldness; women may wear high heels to appear taller. When we notice someone pulling into an accessible parking space, hopping out, and running into the store, seemingly obvious that they are not disabled, we judge them for it. We conclude that they are lazy or inconsiderate, and we may even become angry that they have robbed the spot from someone who actually needs it. Our reaction may be to just shake our heads in disbelief and continue about our business, or our sense of morals may compel us to throw them a disapproving look or even say something to let them know that they have been caught. We want them to know that they are not "pulling the wool over our eyes!"

Not all disabilities are visible. According to Accessibility.com, about 25 percent of adults in the United States have a disability, and most invisible disability metrics in the United States say that the number could be as high as 20 percent or more for Americans with an invisible disability. That said, there is a high likelihood that we have co-workers with invisible disabilities. There are no outward signs of injury, ailment, or impairment, nor do they use assistive devices such as a cane or wheelchair. From our vantage point, they have no medical issues whatsoever. When we encounter someone experiencing severe migraines, asthma, or sickle cell anemia, for example, we are prone to judge them when their health challenges lead to absenteeism, missed deadlines, or the belief that they are not doing their fair share of the work. A person with an invisible disability may seem perfectly healthy yet live with many challenges. Their battle with extreme fatigue or cognitive impairments may go unnoticed, deceiving us into believing that all is well. We are oblivious to incapacitating pain, dizziness, or weakness, which periodically or constantly prevents them from functioning as well as everyone else. Further, our disbelief heightens when we encounter young people experiencing ailments that we associate with aging, like rheumatoid arthritis or chronic back pain. Meet 22-year-old Darryl. He's been a sales associate at a locally owned bookstore for six months. In addition to working with customers, the nature of the work requires accepting and inspecting shipments, conducting and managing inventory, maintaining store cleanliness, and participating in seasonal refreshes of the store signage and displays. Darryl is hard working, is attentive to the needs of customers, and gets along well with co-workers. By all accounts, Darryl is the model employee. As the Christmas season approached, it was time to do a major refresh to include retrieving stored items, assembling and hanging decorations, and moving furniture. The team of four was working late to start the process. Everybody had a job to do. Darryl, who stood 6 feet 2 inches and weighed about 225 pounds, was assigned the role of retrieving boxes from storage. When he respectfully declined the task citing chronic back pain from a sports injury, no one took him seriously. His weight, height, and consistent pleasant demeanor hid his condition. One co-worker expressed utter disbelief when she announced, "Oh, all of a sudden, your back hurts at a time when we rely on you the most!" Even the store owner was shocked as Darryl had never disclosed his condition. In that moment, Darryl felt his character and integrity under attack. If only they knew how much he wanted the same thing, the ability to live life on his terms—not the terms his back pain imposed. Usually, we don't spot these challenges and may find it difficult to understand or believe that someone with a hidden impairment genuinely needs support. We believe what we see, and if it can't be seen, it simply does not exist.

Invisible disabilities include mental, physical, or neurological conditions that go undetected but can impact a person's movement, senses, or activity. Managing an invisible disability is often accompanied by the stress of hiding it. Many suffer in silence rather than disclose the condition for fear of being judged, labeled, or stigmatized. Why risk the possibility that a co-worker may laugh in disbelief, turn in disgust, or diminish the disclosure, when what is needed is to be understood? Many feel that they are not supported at work and feel pressured to prove ailments are real. In a post on BBC.com, Nessa Corkery shares that during her first years of school, she was diagnosed with dyslexia. Mayo Clinic describes the condition as a learning disorder that involves difficulty reading due to problems identifying speech sounds and learning how they relate to letters and words (decoding). As an adult, she studied nursing and landed a position in a hospital. She explains, "I have always been a very confident person and hate to let people think that just because my brain processes things differently that I am not able to do what others can do." The support available while in school, which helped her succeed, fell short in the workplace. Eventually she realized that she was not able to keep up no matter how hard she tried. She states, "It looked like I was complacent and just didn't care, but I just found it difficult to retain everything at the pace the others could. Nursing staffing shortages are a constant issue, so staff nurses are usually too stressed to take the time to teach students. I found it very difficult to ask for extra help as I was already considered a hindrance." One of their greatest difficulties is managing the pressure of getting things done as expected, whether it's an internal self-imposed pressure or pressure felt from colleagues. To be able to say, "I just can't do this today," or ask for help would go a long way in creating a sense of belonging and inclusion. Our lack of sensitivity to someone's disability, especially an invisible one, can lead to misunderstandings, resentment, and frustration. It is impossible for us to be empathetic until we have acknowledged there is a situation for which to be empathetic. If someone uses a prosthetic limb or is visually impaired, it can be easier to understand the difficulties they may face and support them. For others where impairments are not visible, such as depression, multiple sclerosis, or epilepsy, it's often a different story. Thus, it's best to avoid judging and determining ability by what we observe. In addition, we need to be mindful of our language. Everyday terms such as *retarded, crazy, nuts,* and *psycho* that we use to describe dislike can diminish the work experience of colleagues and should be avoided. We are ignorant to the harm we are causing and can substitute these terms with alternatives such as *outrageous, absurd,* or *illogical.* Our inclusion journey requires that we seek what lies beneath the surface, despite what we see on the outside.

NOTE   Note that the people depicted in this image are models and are not in actuality associated with any particular illness.

Figure 18.1: Can you tell who has which disability?

Actions

Resist the Temptation to Make a Visual Diagnosis

Visual observations tell only part of the story. Champions of inclusion strive to learn more. Examine types of invisible disabilities. See Figure 18.1 to get started. Get to know the psychological impact as well as challenges and limitations. Choose not to judge based on appearances. Decide now how you will respond when you become aware that a disability exists.

Make Support Ongoing

Everyone, regardless of ability, requires the support of others to be successful. Rather than ignoring or responding in disbelief to the revelation that a co-worker lives with an impairment, be supportive. Listen more than you talk, and step in or step up when a colleague is discriminated against or disrespected (see Activity 26, "If You See Something, Say Something.") Join or create an employee resource group that focuses on disability in an empowering forum with allies to network and raise issues.

Consider a Day in the Life

Inhabit the shoes of people with invisible disabilities by listening to their stories. Check out:

- **YouTube.com:** *The Hell of Chronic Illness*, Sita Gaia, TEDxStanleyPark: www.youtube.com/watch?v=sKtbhZpTpbc
- **YouTube.com:** *Confronting the Invisible*, Olivia Larner, TEDxFurmanU: www.youtube.com/watch?v=4QyUdMFNb6s

Simulations of invisible disabilities:

- **YouTube.com:** *The Party: a virtual experience of autism*, 360 film, posted by *The Guardian*: www.youtube.com/watch?v=OtwOz1GVkDg
- **YouTube.com:** *I Wanna Go Home*, VR Intellectual Disability Simulation, Enliven, Social Enterprise: www.youtube.com/watch?v=YwnHfdXbras

Action Accelerators

- **YouTube.com:** *Invisible Disabilities*: www.youtube.com/watch?v=f9OEEU90qds
- **MindPathCare.com:** "Do you know when you're using harmful ableist language?": www.mindpath.com/resource/do-you-know-when-youre-using-harmful-ableist-language
- **Cdrnys.org:** "Did You Know? Invisible Disabilities," Center for Disability Rights: www.cdrnys.org/blog/development/did-you-know-invisible-disabilities

- ■ YouTube.com: *Sherpas: Climbing the Mountain of Bi-Polar*, Debbie Foster, TEDxCrestmoorParkWomen: www.youtube.com/watch? v=YnihCgrsz3s

- ■ Forbes.com: "Ableism in the Workplace: When Trying Harder Doesn't Work," by Nancy Doyle: www.forbes.com/sites/drnan cydoyle/2019/11/24/ableism-in-the-workplace-when-trying-harder-doesnt-work/?sh=3a29b41915ae

Sources Cited

Mauro Galluzzo. "We Need to Talk about Dyslexia at Work," BBC.com, July 2, 2019, www.bbc.com/worklife/article/20190 702-we-need-to-talk-about-dyslexia-at-work

Invest in the Success of Others

"An individual has not started living until he can rise above the narrow confines of his individualistic concerns to the broader concerns of all humanity."

—**Martin Luther King Jr.**

***

As more organizations focus on diversity, equity, and inclusion, leaders are setting goals to not only diversify the workforce as a whole but also its leadership ranks. The journey is fraught with pitfalls when they expect to achieve these goals while maintaining existing practices. Looking around your organization, you may notice that the bulk of employees of difference are at entry-level ranks, a few may have achieved middle manager positions, and there are little to no Black or Brown people at the most senior levels. The Civil Rights Act of 1964 made it illegal for employers in the United States to discriminate based on race. Despite legislation forbidding racial discrimination and overt acts of prejudice, systemic inequalities remain, and racist attitudes have become more subtle, while some in the dominant culture doubt that racism still exists—after all, we elected a Black president. However, leaders cite a multitude of reasons for their mostly homogenous workforces. The ones I hear most often include, "People of color don't apply," "We can't find *qualified* diverse candidates," and "I hire only the best person for the job," signaling that the bar is lowered for diverse applicants. A post from CNN.com shares, "Wells Fargo CEO Charlie Scharf has apologized after he blamed the lack of diversity at the bank on a very limited pool of Black talent to recruit from." The apology came in a memo to employees released by the bank for "an insensitive comment reflecting my own unconscious bias."

"There are many talented diverse individuals working at Wells Fargo (WFC) and throughout the financial services industry, and I never meant to imply otherwise," he wrote. "It's clear to me that, across the industry, we have not done enough to improve diversity, especially at senior leadership levels. And there is no question Wells Fargo has to make meaningful progress to increase diverse representation."

If the human resources (HR) team, hiring managers, and senior executives were to dig a little deeper in these instances, they would realize that they have a process problem, not a lack-of-qualified-candidates of color problem. The process of attracting, recruiting, and ultimately hiring members of historically excluded groups requires an intentional process. However, diversity recruitment strategies will not be addressed in this book but bear being called out as a catalyst for the dilemma. This backdrop gives context to champions of inclusion. The typical, everyday champions of inclusion are not necessarily those who make hiring decisions or change recruitment policies and processes but can, however, advance change by investing in the success of co-workers who they interact with. That said, if you are a hiring manger, HR professional, or senior executive, this guidance is useful for you too. While organizations place emphasis on diversity recruitment, retention and advancement get overlooked. Chances are that when an organization struggles to attract diverse talent, they also struggle to keep that talent. A post from KazooHr.com cites an interesting fact from a study by Center for Talent Innovation: "More than one-third of Black employees plan to look for a new job in the next two years. That number is 30 percent higher than it is for their white colleagues." Regardless of our rank in the organization, we have unique skills and competencies that can be invested in someone else. We will reap improved relationships, while creating a more energetic workplace culture that is full of people who are maximizing their potential. Low employee retention rates can occur for any number of reasons, but the most common ones are a toxic workplace culture and limited to no opportunity for advancement. By contrast, a high-retention workplace maintains employees for years as the culture is conducive to their success, personal and company values align, and they are committed to organizational goals.

When Carmen, a senior manager at a technology services firm was asked by two colleagues to join them for a short-term project as part of a company-wide contest, she was intrigued. One colleague, Amber, was a direct peer from a different division whom she knew quite well, and the other, Cassandra, was a junior colleague from yet another division whom she barely knew. The contest came with a $10,000 cash bonus to the winning team. She pondered whether her skills would be of value but decided to join them anyway. She figured that even if they did not win, it would be fun creating something entirely new, and she could spend more time with a colleague she respected and get to know the other better. The team of three worked remotely in different parts of the country and was one of a dozen or so contestants. They had eight weeks to come up with a winning idea and strategy before presenting to the entire organization. By the third week, Carmen realized that she was doing the brunt of the work and started to harbor resentment. She was managing the entire process—scheduling and leading their weekly meetings, building agendas, and sending recaps complete with agreed upon follow-up items and deadlines. She was waiting for one of them to take note of all she was doing and offer to step in to share the load. After the third meeting, her resentment turned to anger, and she knew she had to share her feelings to avoid damaging the relationships. What she discovered during conversation is that the two admired her organization and leadership skills and felt ill-equipped to direct the project as well as she. Cassandra revealed that she lacked confidence and was shy about contributing in meetings for fear of not being taken seriously, and Amber liked the ideas and direction from Carmen and preferred to play a support role to maintain momentum but agreed to rotate meeting logistics duties. This was not the outcome that Carmen was expecting. She had hoped her colleagues would figure out ways to contribute more, perhaps research solutions already in the market, reach out to contacts for additional insight, or look for white papers and studies on the topic as she had done. She was in it to win it. With four weeks left to finalize a solution and a week needed to prepare the presentation, Carmen decided to proceed with the few adjustments they discussed, but at the same time had come to realize her value. Carmen attributed her competencies to the guidance from mentors and advisors she'd had throughout her career and decided that after the project was completed, she would pay it forward. Armed with the revelations from their discussion, she sought links to relevant articles, podcasts, and expert bloggers, as well as a few books, and passed them on. She offered to meet with the two of them every other week for a few months and then as needed after that. They really appreciated her efforts and advice. Over time, Carmen could see the improved confidence in Cassandra. She was a lot more engaged in their discussions and was clearly applying her newly acquired knowledge in her job role. Nine months later, Cassandra was promoted. Carmen felt gratified to think that she played a part in her success and had also gained a solid relationship in the process. Amber lost interest after a few months but continued to not only reach out for support from time to time but also offered support, something that she had not done previously.

It's not unusual for teams to have diamonds in the rough—those high-potential individuals who embody teamwork, have a strong work ethic, or exceed expectations over and over with little recognition. When these extraordinary individuals happen to be women and members of underrepresented groups, they can be overlooked as such and may become disillusioned, thinking that performance alone is the key to success and advancement. While we should make an effort to be supportive of all colleagues, taking extra steps to invest in the success of another whom you believe to be exceptional is a rewarding experience, makes them feel more included, and positions them to thrive. Discuss career goals and aspirations with colleagues who are different from you. Collaborate on career strategies. Share best practices and unwritten rules. Consider a mentorship, sponsorship, coach, or advisor role. Your investment will be a win for you, a win for them, and a win for the organization. The trifecta!

Actions

Help Others Achieve Success

Potential is all around us. Take note of who is getting promoted or supported. Discern whether women and members of other underrepresented groups are included. If they are, great! Seek opportunities to add more value. Where you see inequities, consider how you can lend your expertise to change the career trajectory of at least one person.

Reach 'Em While They're Young

Often members of historically excluded groups are not exposed to role models. You may have heard the saying, "You can't be what you can't see." Volunteer at an organization that prepares members of underrepresented groups for the workforce, i.e., Junior Achievement, iMentor, or Girls Who Code.

Increase Exposure—Create Opportunity

Advocate for women and members of other underrepresented groups as keynote speakers or panelists at company or industry events.

Recognize individuals in nonleadership roles for their leadership abilities by talking up their competencies and accomplishments to others, especially those with positional authority. Seek opportunities to speak of a job well done.

Action Accelerators

- **YouTube.com**: *On Diversity: Access Ain't Inclusion*, Anthony Jack, TEDxCambridge: www.youtube.com/watch?v=j7w2Gv7ueOc

- **YouTube.com**: *Why diversity initiatives fail*, Khalia Newell, TEDxCityUniversityLondon: www.youtube.com/watch?v=Dl0QRaeyTf4

- **Forbes.com**: "Invest In High Potential Women. It Is Important," by J Alaina Percival: www.forbes.com/sites/alainapercival/2020/07/30/invest-in-high-potential-women-it-is-important/?sh=2975d9071b89

- **JPmorganChase.com**: "The Importance of Mentors and Sponsors in Career Development": www.jpmorganchase.com/news-stories/the-importance-of-mentors-and-sponsors-in-career-development

- **Inc.com**: "3 Things to Cover During Your Mentoring Sessions," by David Finkel: www.inc.com/david-finkel/3-things-to-cover-during-your-mentoring-sessions.html

Sources Cited

Chris Isidore and Matt Egan. "Wells Fargo CEO Apologizes for Saying the Black Talent pool is Limited," CNN Business, September23,2020, www.cnn.com/2020/09/23/business/wells-fargo-ceo-bias/index.html

KazooHR.com. "10 Strategies to Retain Diverse Talent (Employee Turnover+DE&I),"www.kazoohr.com/resources/library/talent-retention-diversity

Lead Change One Word at a Time

"Those whose business it is to open doors, so often mistake and shut them."

—George MacDonald

***

Speaking is part of everyday life. Our choice of words exemplifies beliefs, values, and feelings. How others interpret our language not only opens the door to better relationships and strengthened connections but also, by contrast, creates barriers and negatively impacts someone's sense of belonging. When language is used for good, we can envision a tomorrow that won't be like yesterday as we use words to change minds and lives, motivate others to action, and demonstrate acceptance. The terms we deploy express what is in our hearts and form how others experience us in conversation. They can encourage and empower those who have been historically marginalized or disenfranchised to triumph over adversity. On the other hand, the words and phrases we unconsciously choose and combine can unwittingly reflect our stereotypical views and biases. Our biases hide in the shadows. When used carelessly, our words can express feelings of our own superiority or our beliefs of the inferiority of certain groups or types of people. For example, a simple statement like "Women are just as good as men in data science" subtly perpetuates the speaker's sexist view that men are better at data science than women. While the intention was to convey that both men and women are equal in their abilities, the sentence structure compares women to men and implies that men being good in data science is the norm. A slight change in wording positions women and men equitably. Saying, "Women and men are equally competent in data science" conveys the intended message. Or suppose two male colleagues, each married, are engaged in small talk about their weekend activities with family when one says, "How did your wife enjoy the venue?" The other responds, "Oh, my husband thought it was great." The wife question reflected the speaker's belief that marriage consists of only a husband and a wife. In addition, the correction probably came with feelings of awkwardness for both men in varying degrees. Had the speaker been aware of how traditional beliefs show up in conversation and may potentially offend or exclude, they could have said, "How did your spouse enjoy the venue?" Sometimes, it is our reaction to information that can reveal bias. Consider, for instance, chatting with a Latina colleague over lunch at a Mexican restaurant. You think it would be fun to place your order in Spanish, so you ask her for help. Your expression turns to shock when she explains that she does not speak Spanish. The reaction reflects a cultural assumption of LatinX individuals, an underlying belief that speaking Spanish is a prerequisite for being Latina. Your colleague may feel judged as you have now deemed her an inauthentic Latina for not speaking Spanish. While the words in the question did not imply bias, the reaction did. When Henrietta, a woman in her early fifties brought her seven-year-old daughter to the office for "Bring Your Kids to Work Day," she was complemented several times on how much her granddaughter looked just like her. She was embarrassed the first few times and politely responded, "I don't have any grandchildren." She wondered why people instantly assumed that grandmother was the only option. We must continuously challenge what we believe about those who are different from us. It's how we learn to value and appreciate difference. These are examples of honest yet avoidable mistakes. The only real mistakes are the ones we don't learn from. Each interaction may create new opportunities to learn something new and help us become more aware of how our biases are reflected in our words, actions, or reactions. Checking assumptions through the lens of inclusion can mitigate biased language that demeans or excludes people based on race, age, gender, sexual orientation, ethnicity, social class, or physical or mental traits.

Even the slightest differences in word choice can correspond with biased beliefs and introduce antiquated terms and norms that were once acceptable. We need to use words carefully and stay current with accepted usage. Champions of inclusion work to create a workplace culture of awareness and sensitivity while striving to be conscientious of terminology that people may find offensive. It's part of our ongoing journey. With today's workforce consisting of at least four generations, each with phrases and terms representative of their era, interactions between co-workers of different generations can easily lead to misunderstandings. Terms that are commonplace for one generation can differ drastically in another. Language evolves regularly to adapt to new mindsets, beliefs, and cultural shifts. You may discover that words you've used for decades without incident are inappropriate or are now considered offensive—for example, referring to Asians as *yellow* or *oriental,* using *homosexual* to describe members of the LGBTQ community, or saying *whippersnapper* when referring to young people. While many accept that language is fluid and impacted by social change–others resist it. It's only natural to stick to the tried and true. Without raised consciousness of the impact of the tried and true, we'll continue to make regular deposits to cultures of exclusion.

Most common offenses are related to gender, race, sexual orientation, or physical traits. We must make sure that we're keeping up. Pay attention to everyday words and phrases that reflect bias, cause harm, or conjure negative reactions. Remove them from your vocabulary and swap exclusive terms and phrases for inclusive ones. Be open to shifts in language and learning from one another. The key here is to have an open mind, challenge assumptions, and work toward speaking more inclusively. This will allow us to build relationships as we improve communications across generations. When we have strong relationships, misunderstandings are less common, and when they do happen, we are less likely to attribute them to malicious intent. Language is a very powerful tool for change. Use it for good. You're leading the charge against division and separation one word at a time. Refer to Table 20.1 to get started.

Table 20.1: Inclusive Language—Say This, Not That

SAY THIS	NOT THAT
Gender	
Women, woman, girls	Gal, gals, females
Transgender	Transgendered, transman, transwoman

Continued

Table 20.1 (*continued*)

SAY THIS	NOT THAT
Full figure (referring to a woman's weight)	Fat, overweight
Flight attendant	Stewardess
Server	Waiter, waitress
I would like to speak to the manager. Are they available?	I would like to speak to the manager. Is he available?
Good morning. Hello, team. Good day, all	Hello, guys. Ladies and gentlemen (when addressing groups of mixed gender and gender identities)
Administrative assistant, admin, executive assistant	Secretary, woman secretary, male secretary
Ability	
Impaired	Handicapped
They use a wheelchair	They are confined to a wheelchair
That's irrational, unreasonable (descriptors of things that don't make sense)	That's retarded, psycho
Person with autism, person with name of invisible disability)	Autistic person
Person living with dyslexia (or other impairment)	Stricken with dyslexia, suffers from dyslexia
Race	
I like your (new) hairstyle. (If it's sincere. Otherwise, say nothing, and don't stare)	Can I touch your hair? How long did that take? Is that your real hair?
Asian, Asian American	Exotic, exotic-looking
	Yellow, oriental (antiquated terms)
Black, African American	Colored, negro (antiquated terms)
Multiracial, biracial	Mixed race, half breed, mutt, Mulatto (antiquated term)
Native American, indigenous	Indian
	Red, redskin (antiquated terms)
That's so inappropriate, unkempt, obnoxious, inferior quality (accurate descriptors)	That's so ghetto (comparing societal norms to lifestyle and behaviors of people of color)

SAY THIS	NOT THAT
Black lives matter	All lives matter. (Indeed they do, but the statement diminishes the racist and prejudicial experiences of Black people.)
LGBTQ+	
That's so cruel, illogical, insensitive (accurate descriptors of the situation)	That's so gay
Husband, wife, spouse (if married) Partner (if dating)	Domestic partner, boyfriend, girlfriend
Gay, lesbian, trans, nonbinary	Homosexual, homo
Age	
Older, 40+, 50+, etc	Dinosaur, relic, old lady, old man
Short lapse of memory, distracted	Senior moment, brain fart
Youthful	Fresh meat, fresh face, new blood
Promising, potential	Inexperienced, not a fit

Actions

Get to Know the Multigenerational Workforce

Expand your network to include all generations, and learn to navigate generational divides for increased understanding.

Become aware of age-related biases. Is someone too young for this or too old for that? Set in their ways? Investigate those types of beliefs and whether they are indicative of the individual you have in mind. Adjust accordingly. You may discover those beliefs to be unfounded.

Mind Your Words

Reflect on popular words and phrases that you've used throughout your life, and consider the impact and relevance in today's workplace. Ask yourself whether you are unintentionally causing harm to others with continued use.

Just as you would check written communications for spelling errors, examine written communications for biased, offensive, and antiquated terminology before clicking 'send'.

During conversations, consider the whole of the individual in front of you. Challenge assumptions and what you believe to be true about them, and put those assumptions in check. Listen and be mindful not to judge. When you make a mistake, apologize, and learn from it.

Action Accelerators

- **zapier.com**: "The Conscious Style Guide: How to Talk About People with Inclusive and Tactful Language," by Genevieve Colman: https://zapier.com/blog/communicate-inclusion-and-diversity

- **Babbel.com**: "Baby Boomers and Millennials: Are They Even Speaking the Same Language," by Thomas Moore Devlin: www.babbel.com/en/magazine/are-millennials-killing-language

- **YouTube.com**: *How generational stereotypes hold us back at work*, Leah Georges: www.youtube.com/watch?v=dKNu5ZnWhb4

- **YouTube.com**: *Cultivating a culture of inclusion*, Life at HSBC: www.youtube.com/watch?v=oSO3F9z23vU

Practice Common Courtesy

"The truest form of love is how you behave toward someone, not how you feel about them."

—Steve Hall

***

What comes to mind when you think of common courtesy? Perhaps you were raised in a household that taught the importance of saying please, thank you, and sorry. Other words to live by may have included respect your elders, speak when spoken to, and hold or open a door for the next person. Maybe you were raised to treat all with dignity and respect. It could be that for you, common courtesy means treating people the way you would want to be treated. As adults, we don't hear as much about common courtesy as we did growing up. In the workplace, common courtesy is a must-have and extends itself to professional courtesy. Common courtesy demonstrates acts of kindness on a human level as mentioned in the aforementioned examples, while professional courtesy demonstrates acts of mutual respect for co-workers. It comes across in many ways to include being attentive in meetings, waiting for your turn to speak in conversations, and giving credit where credit is due. We should practice common and professional courtesy as they symbolize respect. Respect is one of the most critical factors in the workplace as it supports an environment of collaboration whereby teams can accomplish goals together, regardless of how they feel about one another. However, being kind and polite to others doesn't diminish us in the workplace, but rather bolsters our reputation and demonstrates our character. While we may be highly competent in our roles, it's how we treat others that drives inclusion. When we combine common and professional courtesy with respect, we empower ourselves to connect across disparate communication styles. Common and professional courtesy is often unspoken amongst co-workers until someone has violated its principle. It's only written about in the employee handbook in the code of conduct section. It's what we've come to expect but, somehow, don't always get. Sometimes it may be hard to find it at any level within the organization as many have adopted the "I gotta be me" mantra and are proud of it. Everybody has or has had at least one co-worker who is intentional about being respected more so than liked. That's the a-hole that most of us choose to avoid. If we're lucky, we may find courtesy and respect in pockets of various teams or catch it in a flash of fleeting kindness. The values of common courtesy, professional courtesy, and respect are not dependent on context. There is not a single instance or circumstance in the workplace where courtesy and respect should not be the default. Depending on how far we drift from it, people may experience job loss, and companies may experience litigation. The further we drift, the more toxic the environment becomes.

We feel better connected at work through positive interactions based on these values as they are powerful contributors to our sense of belonging. In essence, we are creating the conditions that will enable all co-workers, but especially those from historically marginalized groups, to thrive at work as stereotypical beliefs, prejudice, and bias may erode courtesy and respect. When these values are present, our sense of well-being and relationships flourish; and in their absence, we become vulnerable to isolation and feelings of self-consciousness. Imagine that you're about to pass a co-worker in the corridor. You attempt to make eye contact, plan to greet them with a smile, and perhaps even say something as you approach. When you get closer, they avoid eye contact, and when you say, "Good morning," they proceed right past you without breaking stride or uttering a sound. It's as if you didn't exist. You're certain that they saw you, you're certain that you were heard, so how could they not respond? You were just inches apart at the moment of passing.

This has happened to me more times than I can count in the workplace. Each time, I felt invisible and pondered what I did to deserve such rudeness. As a young professional, I blamed myself for the longest time. I protected myself by never initiating a greeting with those individuals again and was always uncomfortable in their presence going forward. The feeling of being invisible never went away. Thank goodness that I've evolved over the years to realize that it wasn't me at all, but them.

Genuine greetings are just one of many small gestures within the courtesy and respect space that make people feel seen and valued. How do we as individuals benefit by avoiding such small gestures? Sure, there are cultural differences in relation to personal space and eye contact that we should consider, but I would like to think that courtesy and respect are universal. When people feel valued, they often value others. It's the gift that keeps on giving and can permeate throughout the organization while exposing the company's bad apples to be either upskilled or weeded out. Champions of inclusion strive to normalize common courtesy and respect through example, apologize when we get it wrong, and accept it as a teaching moment to do better in the future.

Actions

Make Professionalism a Habit

Follow through on your commitments and responsibilities. Be a person of your word. Colleagues will respect you more when they know that they can depend on you.

Help co-workers when they are struggling. Respectfully offer advice or assist with a portion of their work to eliminate stress. Encourage them when they fail and provide guidance on how to do better in the future.

Think Before You Speak

Words have power (see Activity 20, "Lead Change One Word at a Time.") Treating people with respect and courtesy comes across in the words you choose and the tone in which you deliver those words. Be aware of your words and how they might affect co-workers. Ask colleagues, peers, and co-workers for feedback. You may be surprised to discover how you are received. After you receive feedback prepare to try something new. Alter behaviors to promote personal growth and the well-being of others.

Express Gratitude

Show appreciation for common and professional courtesy. Send hand-written notes of thanks, and publicly acknowledge a job well done or when someone goes the extra mile. Praise accomplishments and achievements in public; provide constructive feedback and criticism in private. Reciprocate favors. Create a kudos Slack channel where everyone can get involved. Gratitude is contagious; repeat often.

Meet and Greet with Authenticity

There's plenty of opportunity to share a genuine greeting. Arrive a few minutes early to virtual or in-person meetings for small talk. Acknowledge folks at water coolers and elevator banks rather than staring at your phone. Consider the last time that you shared a simple hello with a co-worker with whom you seldom interact. It's probably been a while. Go ahead and make their day.

Greet people by name, and pronounce it correctly.

Self-Reflect

Courtesy and respect in communication is not only about what we say, it's also how we say it. Consider tone of voice and body language when interacting with others. Reflect on whether your demeanor is saying one thing and words another. Examine whether this effect occurs often and with only certain individuals or groups. Be honest with yourself.

Action Accelerators

- PsychologyToday.com: "The Power of Hello: The simplest way to make your world a better place," by Sam Sommers: www.psychologytoday.com/us/blog/science-small-talk/201203/the-power-hello

- Inc.com: "Why You Should Greet Your Coworkers Every Day," by Lindsay Dodgson: www.inc.com/business-insider/why-you-should-greet-your-co-workers-everyday.html

- **Shiftworkplace.com**: "The spirit of work: How courtesy, respect and thoughtfulness create the foundation for all other work success": shiftworkplace.com/spirit-work-courtesy-respect-thoughtfulness-create-foundation-work-success

- **DiversityAndrespect.org**: "The 30 Tips of Dignity & Respect": dignityandrespect.org/the-30-tips-of-dignity-respect

22

Amplify Voices That Aren't Being Heard

"The function of freedom is to free someone else."

—Toni Morrison

***

We are empowered to perform our best in the workplace when we feel that our voices are heard. We need to know that our contributions are valued, appreciated and, when we're on the money, that our ideas will be explored or challenged by peers and colleagues. When we feel ignored, it is as if our experience, skills, and competencies are of no merit. As a result, we feel excluded, and as anyone who has ever experienced this knows, it impacts our morale and well-being. We want others to listen in such a way that we know we've been understood even when we agree to disagree. Effective communication occurs when we acknowledge what the other is saying, even though we may not like what's being said. In the name of respect, individuals need to be allowed to speak without being interrupted, discounted, or having their message discounted, especially on the basis of difference. All too often members of marginalized groups are not heard, because the dominant group feels that they know "how it's done." Women are not heard, because men need to be right or feel that women are underqualified on the subject matter and subsequently take over the conversation with mansplaining. According to Merriam-Webster, *mansplaining* is when a man talks condescendingly to a woman about something he has incomplete knowledge of, with the mistaken assumption that he knows more about it than the person he's talking to. When people go unheard time and again, eventually resentment builds, and withdrawal occurs. After a while, efforts to communicate are abandoned, and the team as well as the organization has lost the benefit of diverse perspectives.

My journey from there to here has included experiences of me sharing ideas in meetings that were met with blank stares. I could not tell whether colleagues were even listening or intentionally chose not to respond. The room would be dead silent for a few very long and uncomfortable seconds until someone spoke up to move on to another topic or back to the previous one. At that point, anything would have been nice—a smile, a nod, something to validate that I was actually in the room. My ideas were glossed over at best or totally ignored. I can attribute it only to the fact that I was the only woman on the sales team, the only person of color on the team, and working in the technology industry, which is dominated by white men. Needless to say, I most definitely did not feel heard. I just kept shaking it off. Eventually I stopped contributing and kept my head down to continue to make quota.

Meet Yuen. It's Monday morning, and she has another full day of back-to-back internal meetings as a financial analyst. Yuen has been in the industry for five years and at her current company for six months. Mondays were bad enough as it is, but one meeting after another is nothing to look forward to when you are someone who seldom gets a word in. Often inclusion is left just outside the meeting room door as the dominant group leaves little to no space for others to contribute. Yuen was diligent about reading the agendas in advance and came prepared with questions and talking points. She was often interrupted mid-sentence by someone in the room while responding to a question or leading with a key point. She was accustomed to pushback and never took it personally. She enjoyed a healthy debate and defending her ideas. However, she could never get that far on her current team. The frustration was becoming overwhelming. She wasn't being listened to, and felt sidelined as a team member. One afternoon as she worked up the courage to share her thoughts and was again cut off, her emotions got the best of her. She burst into tears and excused herself from the room. She attempted to calm down by walking around the beautiful corporate campus, but every time she thought that she was ready to walk back into the office, the tears came rushing back. She decided to take the rest of the day off to regroup. Upon her return on the following day, she decided to meet with her manager for guidance on how she could better contribute. He was an industry veteran and a well-respected man in his mid-50s. She was confident that his wealth of knowledge would deliver sage advice. Without an appointment, she popped into his office and asked if he had time to talk. He smiled and welcomed her in. She began by telling him how greatly she appreciated the opportunity to be at the firm and started explaining her challenges since her arrival. As she shared her experiences, the emotions started to resurface. She fought back the tears and continued. He was seemingly listening. He was making eye contact and even nodded here and there. She thought, finally, I'm being heard. When she finished, she waited patiently for his feedback. Instead, he responded, "You're doing just fine. Keep up the good work." He then excused her claiming that it was time for his next appointment.

One of the most powerful ways to enhance any working relationship is to value individuals and interactions over processes and take the time to actively listen during conversation. Active listening is described as the practice of listening attentively to the speaker for understanding and conveying that you've absorbed the message. This benefits both participants as they continually check for clarity. A post on LinkedIn cites, "Statistics highlight that for approximately 80 percent of our day, we all are engaged in communication. Out of this, about 55 percent is spent on listening to others in our surroundings." The post goes on to invite readers to ponder whether they are active or passive listeners and details best practices. See sources cited for details.

Champions of inclusion work to shift the tide and amplify the voices of those who are not heard. The key to success is self-awareness and moving personally from a passive or selective listener to an active one. We need to ensure that we are not part of the problem. In addition, we must determine what we are comfortable doing and always be prepared to carry it out. When we see evidence of our input being taken seriously, it motivates us to do more and grow professionally. It's so much better than a mere pat on the back or a half-hearted promise to "take that under consideration." We can foster a more equitable workplace where everyone can be heard.

Actions

Build Inclusive Meetings

Group dynamic in meetings is probably the most common situation where people can be shut off from contributing. At your next meeting, take note of who is doing the talking, who is being talked over, whose ideas are being explored, and whose are ignored. Look for opportunities to do the following:

- Interrupt the interrupters with a few prepared phrases like: "I think Susan has something to say" or "Please let Armando finish his thought. I would like to hear it."

- Open the door for the rightful owner to take credit for their idea when someone else claims it. Say something like, "Kenyatta's idea was very similar; now may be a good time for her to expand on it."

- Examine whether there is a pattern of the same individuals being heard while others are forced to sit on the sidelines. Those with type-A personalities tend to be the ones who dominate most conversations. They are characterized as being ambitious, competitive, impatient, and driven. Consider having a sidebar with those who take over discussions and explore their openness to speaking last on ideas presented to ensure that all who want to contribute can do so. Use a collaborative tone, let them know what you've observed, and explain how it affects the team dynamic and goals to help them see the importance of making the shift. (For more on personality types, see the link in "Action Accelerators.")

- Check the room from time to time for signs that someone may be trying to say something but has not been able to jump in. Invite them to speak.

Empower Others to Amplify Their Own Voices

We can equip others to amplify their excellence by providing tools and strategies that build confidence and position them for success. Here are a few that you can review and pass on:

- **IrishTimes.com**: "How to Interrupt the Interrupter": www.irish
  times.com/business/work/how-to-interrupt-the-interrupter-
  1.2993632

- **SwitchTheFuture.com**: "Expert Advice for When You're Ignored
  in Meetings": switchthefuture.com/2020/03/31/ignored-in-
  meetings-expert-advice-on-what-to-do

Explore Reasons That People Remain Silent

Not everyone is comfortable speaking publicly, regardless of who is
in the room and the number of people in the room—it's probably just
not their nature. Maybe they have a fear of being judged or will speak
only when invited to do so. If this is the case, encourage them to send a
follow-up email or Slack message to the group to share their ideas. We
have to meet people where they are.

Seek Accomplices

No need to be an army of one. Build strength in numbers. Connect with
a few trusted colleagues to share the load of inclusive practices during
meetings. Create a plan together.

Collaborate with your manager on inclusive meeting strategies and
encourage them to share with other managers.

Action Accelerators

- **HBR.org**: "To Build an Inclusive Culture, Start with Inclusive
  Meetings," by Kathryn Heath and Brenda F. Wensil: hbr
  .org/2019/09/to-build-an-inclusive-culture-start-with-
  inclusive-meetings

- **Atlassian.com**: "Team Playbook: Inclusive Meetings": www.atlassian
  .com/team-playbook/plays/inclusive-meetings

- **HireSuccess.com**: "Understanding the 4 Personality Types: A, B,
  C, and D": www.hiresuccess.com/help/understanding-the-
  4-personality-types

- **FrameShiftConsulting.com**: "Meeting Skills" (download the PDF
  Meeting Role Cards): frameshiftconsulting.com/resources/
  meeting-skills

Sources Cited

Gurleen Kaur. "Active Listening and Workplace Success," April 19, 2022, www.linkedin.com/pulse/active-listening-workplace-success-gurleen-kaur

Engage Remote Colleagues with Intention

"You have to act as if it were possible to radically transform the world. And you have to do it all the time."

—**Angela Davis**

***

As the covid-19 pandemic was in full swing, many employers were forced to embrace the remote workforce as a means to stay afloat. While we are slowly getting back into the office at varying degrees, the remote workforce is here to stay. A post from Apollo Technical shared that "Upwork Global Inc. estimates that 22 percent of the workforce, 36.2 million Americans will work remotely by 2025," and continues that "10,000 employees surveyed by the Becker Friedman Institute for Economics at the University of Chicago said they thought they were just as productive working from home compared to working in the office. In fact, 30 percent of those respondents told researchers they were more productive and engaged working from home." In addition, they report, "Globally, 16 percent of companies are fully remote according to an Owl labs study." The pivot to a remote workforce has widened the candidate pool for employers looking to attract and hire more people of color and those with physical and cognitive impairments. The work-from-anywhere business model affords employers the ability to hire individuals anywhere in the world without ever entering an office. The opportunity is even greater for employers in predominantly White areas who are hoping to leverage the benefits of a diverse workforce as the likelihood of persuading individuals to leave friends, families, and communities behind for surroundings where they may not be appreciated is unrealistic. It's just human nature to want to live, shop, and enjoy leisure activities alongside people with whom we share are similarities.

While access to new talent is good news for employers, this may unintentionally reinforce or exacerbate existing racial and gender hierarchies for those working off-site. One of the challenges faced by historically excluded groups is being visible, not only to colleagues and peers with whom we need connections, but to the leaders who can impact an individual's career trajectory. Out of sight could mean out of mind when coveted projects or opportunities for advancement are on the horizon. When employees are remote, the ability to be listened to and valued for input and contributions is now at greater risk. Visibility in the workplace and the ability to be fully and accurately seen by others is critical to delivering the full force of lifelong experiences toward organizational and team goals. Generally, isolation makes it more difficult for employees to feel connected, build relationships, and feel part of the company culture. While those in physical workspaces benefit from impromptu interactions with co-workers at the watercooler, coffee machine, or cafeteria, virtual teammates do not have this luxury. Chance and casual conversations that build deeper connections are no longer part of the day-to-day routine. Therefore, managers and colleagues must be intentional about engaging with those in a virtual environment. We need to raise our awareness of how we include or exclude teammates during online meetings, create virtual opportunities that support belonging, and otherwise create new habits that bring people into the mix. Our sense of belonging is where we feel safe and valued for what makes us different and allows us to be our authentic selves. We are happier and more productive when we feel that we belong and are better able to deal with job-related stress. By contrast, where there is no sense of belonging, people are more inclined to seek a better environment elsewhere.

While visibility is a consistent challenge for people of color, invisibility (going unnoticed and unheard) and hypervisibility (being overly observed and scrutinized) are part of the lived experiences in the workplace and perhaps more so in a remote environment. When invisibility and hyper-visibility are present, they often act simultaneously to burden members of marginalized groups. Shamika, a Black civil engineer, experiences this balancing act firsthand when being unheard in discussions and situations where her voice should be leveraged as an equal contributor but instead often goes unnoticed. She finds herself continuously defending and proving her work under the guise of feedback, while White male colleagues are seemingly never challenged. In addition, she is often called upon to sit in on client meetings when leaders feel they need to demonstrate diversity. She feels that she is not being recognized as an individual making substantial contributions to the organization due to the color of her skin and gets the spotlight only when her skin color can be leveraged for profit. Members of the dominant group must make the effort to learn about hypervisibility, how they perpetuate it, and how to adjust behaviors that allow hypervisibility to thrive.

Other opportunities where we can make a difference for remote workers is to mitigate bias. Unchecked stereotypical beliefs and bias can wreak havoc on one's career and well-being as some managers remain anxious about not being able to monitor team members to ensure productivity, and co-workers' misconceptions about remote colleagues can poison interactions. Work is more scrutinized or more conflict may arise, resulting in a toxic culture without ever setting foot in the office. Bias, whether good or bad, restricts accurate perception, and absent visibility to change perceptions, bias persists. We've got to build workplace cultures based on trust, inclusive leadership, and transparent policies with up-front conversations about organizational expectations that meet the needs of workers in order to fulfill those expectations.

Ensuring that inclusion is part of the day-to-day for remote workers fosters belonging and helps them to perform at their best. To be effective as well as engaged, they must be both heard and seen. Champions of inclusion continually seek opportunities to connect and engage on a human level, getting to know remote co-workers not only professionally but personally (see Figure 23.1). We all need human interaction, even if it's just through video. My efforts have helped me to meet the spouses, children, and pets of remote colleagues that I probably would never meet otherwise and enables me to better support them through the unexpected while improving our ability to collaborate as team members.

Figure 23.1: Genuinely connect with remote colleagues

Actions

Develop Trust and Connection

Spending time on nonwork activities with co-workers strengthens working relationships. Create a personal connection by first sharing a fun fact about yourself, and invite them to do the same. As the relationships build, ask open-ended questions on topics they care about. It shows that you've been listening.

Create space in your week to interact with remote colleagues. Schedule virtual coffee or lunch breaks. Consider a "happy half hour" get-together at the end of a long day or week. It's great way to catch up. Use video if everyone is comfortable being on camera.

Switch up virtual backgrounds reflective of current mood or interests. Took photos and selfies while on vacation? Upload one as a background. When working on my birthday, I use a balloon bouquet as a background. It's always a great conversation starter.

Arrive to online meetings a few minutes early for small talk with other early arrivers.

Encourage as well as recommend remote colleagues for career-enhancing work as they may not be top of mind when opportunities present themselves.

Be Aware of Your Online Etiquette

Reflect on your online etiquette. Do you talk over people? Pronounce colleague's names correctly? Constantly interrupt? Take note of actions that may be exclusive; work on ways to improve. Employ the same practices in other settings as well.

Action Accelerators

- **SHRM.org**: "6 Ways to Foster Inclusion Among Remote Workers," by Roy Mauner: www.shrm.org/hr-today/news/hr-news/Pages/6-Ways-Foster-Inclusion-Among-Remote-Workers.aspx

- **InHerSight.com**: "5 Ways to Turn Small Talk into More Meaningful Interactions," by Shameika Rhymes: www.inhersight.com/blog/culture-and-professionalism/meaningful-small-talk

- **Flexjobs.com**: "How to Build Strong and Meaningful Relationships with a Remote Team: 6 Tips," by Robin Madell: www.flexjobs.com/blog/post/how-remote-workers-can-stay-connected-with-their-team-v2

- **LinkedIn.com**: "Why the Rise of Remote Work May Help Companies Become More Diverse — and More Inclusive (Talent Blog)," by Samantha McLaren: www.linkedin.com/business/talent/blog/talent-acquisition/why-remote-work-may-help-companies-become-more-diverse

24

Adopt Gender-Neutral Terminology

"Diversity is about all of us and about us having to figure out how to walk through this world together."

—Jacqueline Woodson

***

The world around us has evolved over the past few decades, and the use of gender-specific language has become a thing of the past. Language has the power to shape culture, and this is especially true of the language used in the workplace. Inclusive language, which represents all gender possibilities can make a world of difference in creating truly equitable workplaces. Without the implementation of inclusive language, we risk not only alienating certain groups within the organization but also perpetuating harmful workplace environments that continue cycles of oppression for marginalized groups. While phrases like "You guys did a great job" or "The board is seeking a new chairman" seem harmless, they signal to the women on the team that they did not do a great job and that anyone other than a man will not be considered for the chair role. The default to masculine terms represents exclusion of all other possibilities. Expressions like "We need the right man for the job" are commonplace and reflect a deliberate or unconscious bias that leads to men and women being treated differently in the workplace. A shift in words may seem like such a small thing, but when we neglect to do so, we are unwittingly saying, "You don't belong." Employing gender-neutral terms in the workplace contributes to an individual's sense of belonging and happiness. A post in *Harvard Business Review* states, "High belonging was linked to a whopping 56 percent increase in job performance, a 50 percent drop in turnover risk, and a 75 percent reduction in sick days." The post continues that "Employees with higher workplace belonging also showed a 167 percent increase in their employer promoter score (their willingness to recommend their company to others). They also received double the raises, and 18 times more promotions."

We can reflect the reality of all genders by altering our choice of words. The use of updated language reflects a more modern recognition that gender can be nonbinary. Complimentary expressions like "a great demonstration of gamesmanship" or "our team of craftsmen is unrivaled" can be gender neutralized to "a great demonstration of how the game is played" and "our team of craftspeople is unrivaled." It's so common for us to fall back on using the noninclusive term "guys" in the workplace. I get called a "guy" several times a day in various settings. Even after 18 months of raised consciousness, I still slip up from time to time. Although the term is friendly and casual and works well in so many situations, it's indicative of males and not inclusive of people of other genders. As champions of inclusion, we work toward communicating without any assumptions about our listeners. Refer to Table 24.1.

Table 24.1: Gender-neutral terms for the workplace

REPLACE THIS	WITH THIS
Chairman/chairwoman	Chair
Husband/wife	Spouse
Salesman/saleswoman	Salesperson/sales representative
Mankind	Humankind/people
Mother/father/sister/brother/son/daughter	Parent/sibling/child
Anchorman	Anchor
Cameraman	Camera operator
Doorman	Door attendant

REPLACE THIS	WITH THIS
Weatherman	Meteorologist
Manpower	Workforce/human effort
Spokesman	Spokesperson
Waiter/waitress	Server
Ms./Mr./Mrs.	Mx.
Hey, guys (in mixed gender group)	Hey, team/folks
Tech guy	Technician

The workplace is laden with terms that default to masculinity. Consider terms such as *man hours, manpower, man up*—the list of man phrases goes on and on in everyday language. Start paying attention to how often you use *man* terms and work to replace them with gender-neutral terms such as *labor hours* or *human effort*, or just avoid using them altogether. Use this knowledge to develop your gender-neutral vocabulary, i.e., chairperson, best person for the job, or they/them. Replace "Good morning, guys" with "Good morning, all." The goal here is to default to the most unbiased phrases and expressions that *include* rather than exclude, and we can carry this practice into our interactions with customers and clients. See Figure 24.1, and ask yourself whether your words speak to the entire group. The words we choose can help us, and those who read or hear them create a welcoming culture as well as demonstrate respect for all people. Our words can challenge prevailing norms and conventions that serve as barriers to equality for underrepresented groups. By using gender-inclusive language, we communicate that we value equality and can advance social progress for people of all genders in a subtle but consistent way. Get ready. This one is going to take commitment and practice. As with any other new skill, it gets easier over time.

Actions

Increase Awareness of Masculine Terms

Monitor as best you can how often you default to masculine terms in everyday language. Take note of the ones you use most often. Create a list of substitutions that you can get into the habit of using.

Figure 24.1: Do your words speak to the entire group?

Be mindful of being gender neutral when communicating in writing, including emails, instant messaging, presentations, etc.

Help Others Not to Forget

Gender-specific language is etched into the memories of most of us. It's so easy to default to the terms we've used and heard our entire lives. Help yourself and others remember by creating an office or team challenge: each time a co-worker defaults to the use of a masculine term, they get to drop a quarter, a dollar, something into a jar. Donate the collected funds to a charity that supports marginalized groups.

As more co-workers band together in using gender-neutral terms, help each other by calling their attention to the words they've used. It was probably an unconscious slip. Don't call anyone out; you don't want to become the "word police." Talk to them privately in the spirit of raising awareness, and invite them to do the same when you slip up. No judgment here—just working to get better together.

Create a Slack channel of gender-inclusive terminology so everyone can add new terms and share successes. Recognize co-workers in the channel as they make progress.

Action Accelerators

- **Glassdoor.com**: "Gender Neutral Language in the Workplace: Why It Matters," by Alli Smalley: www.glassdoor.com/employers/blog/gender-neutral-language

- **YouTube.com**: *Win Chesson: Why Gender-Inclusive Language Matters*: www.youtube.com/watch?v=12YNrEgKHZY

- **Forbes.com**: "How to Use Gender-Neutral Language, And Why It's Important to Try," by Kim Elsesser: www.forbes.com/sites/kimelsesser/2020/07/08/how-to-use-gender-neutral-language-and-why-its-important-to-try/?sh=7ca5ff0726ba

- **GroundReport.in**: "Important Gender-neutral words you should start using": groundreport.in/important-gender-neutral-words-you-should-start-using

Sources Cited

Evan W. Carr, Andrew Reece, Gabriella Rosen Kellerman, and Alexi Robichaux. The Value of Belonging at Work, HBR.org, December 16, 2019, hbr.org/2019/12/the-value-of-belonging-at-work

25

Foster an Environment
of Trust

"In the end, you have to choose whether or not to trust someone."

—Sophie Kinsella

***

Trust is a significant factor in any relationship—even those in the work-place. Trust or lack thereof dictates our interactions with one another. That small hint of "something is not quite right" is usually an indicator of a lack of trust. Sometimes it's subtle, like when you notice a colleague who is warm and friendly with some co-workers and has a demeanor that's cold and distant in your presence, or the time when your teammate "accidentally" omitted a substantial fact you needed for the success of a presentation, and you were called out on it. Sure. They apologized, but it came across as insincere. Then there's the utter distrust that happens when you discover that a colleague was deliberately less than truthful with information or gave their word and did not follow through, leaving you in a predicament. Where there is a lack of trust, we tend to be more guarded during conversation, suspicious of motives, or worse, avoid conversation altogether unless absolutely necessary. Who wants to be in an environment at least 40 hours per week with folks we can't trust? Not only is it uncomfortable, but it also adds stress that could be avoided should trust exist. Trust in working relationships contributes to inclusive company cultures and our sense of belonging. When we demonstrate that we can be trusted, connections are stronger. Others get the assurance that when misunderstandings occur, the relationship remains intact. The more trust we have, the easier it is for us to be our best selves in the workplace. Trust is a powerful emotion that makes us confidently rely on the integrity or ability of people and organizations. We benefit from the following:

- More psychological safety with colleagues
- Increased comfort and confidence during interactions
- Open and honest conversation
- The reassurance that someone has our back or best interest
- Collaborative conflict resolution
- Ability to admit mistakes without judgment

However, trust must be earned through consistent actions and behaviors based on what we value. If we value humanity, integrity, and ethics, for example, we will demonstrate those beliefs in decisions and day-to-day interactions. They will shine through every time as we offer respect as a sign that we value the worth of people and accept their imperfections in the process; and hope that we receive the same in return. By contrast, if we value ourselves above all others, our actions and behaviors will be driven by the motivation to come out on top by any means necessary in every situation. Constant displays of a lack of empathy and dishonesty chip away at relationships. We see them when co-workers publicly call out mistakes and their impacts, steamroll over the needs of others, or rely on criticism and blame rather than accept responsibility. Working alongside this type of individual destroys working relationships through the following:

- Fear or hesitation to share and communicate
- Feelings of being disrespected
- Unnecessary conflict
- Lack of connection
- Suppressed emotions of anger, anxiety, or hostility

Throughout my career, I have encountered individuals who eroded my trust over time by disregarding my rights, needs, and feelings in the workplace. The mere mention of their name conjured reactions of tension and anxiety. Resentment built toward the leaders who knew of the behavior and did nothing to stop it—or worse, defended it. The recollection of things that co-workers said and did over the years have left scars—some permanent. One of my favorite quotes comes from Maya Angelou, "I have learned that people will forget what you said, people will forget what you did, but people will never forget how you made them feel."

Learning to build trust at work is critical if you're going to be respected, liked, and connected as a teammate or leader. While many individuals will say that they don't come to the workplace to make friends or be liked, I wonder why they prefer the alternative—being isolated and disliked. If that's what one wants for themselves, it's a personal choice; however, the behaviors associated with that choice can negatively impact the well-being of all others with whom they make contact. As champions of inclusion, we are intentional about nurturing relationships, showing a genuine interest in the whole person, and making that a part of our daily routine. No need to be grandiose. A simple open-ended question like, "Tell me how your presentation went last week" or "What did you do over Spring break," demonstrates that you've been paying attention during previous interactions and that you care. When trust is established, we can have more powerful interactions that improve workplace culture. People feel that they belong and can maximize their potential. When we're trusted by our co-workers, we enjoy better collaboration, get along well, understand each other more, and feel supported.

Actions

Self-Assess

While we think we know how we are perceived in the workplace, it never hurts to verify our perceptions versus reality. List the qualities you believe to be top of mind when someone hears your name and whether your decisions, actions, and behaviors are a threat to others' ability to trust you. This comparison may mark areas of improvement. If you remain unsure, explain this exercise to several co-workers, and ask what the first few words are that they think of when they hear your name. Share that the goal is to improve working relationships and request total honesty. Here's to hoping that their descriptors match yours.

Nurture Relationships

As you focus on building or improving trust in the workplace, be mindful of whether your actions would make others feel comfortable relying on you as well as feel confident in your intentions and motivations. You must be genuine, or efforts will backfire.

Model Consistency

Having the ability to predict that someone will act or react in specific ways fosters a sense of security and confidence in others. Being emotionally intelligent plays a key role here and earns a level of credibility over time.

Action Accelerators

- ▪ `YouTube.com`: *How do I build trust at work?* Simon Sinek: `www.youtube.com/watch?v=WmyfDfCc3_0`

- ▪ `YouTube.com`: *How to build (and rebuild) trust,* Frances Frei: `www.youtube.com/watch?v=pVeq-0dIqpk`

- ▪ `LabManager.com`: "Modeling Inclusive Behaviors to Build Trust with Staff," by Donna Kridelbaugh: `www.labmanager.com/leadership-and-staffing/modeling-inclusive-behaviors-to-build-trust-with-staff-26135`

- ▪ `Medium.com`: "Building and Maintaining Trust in the Workplace," by Paul Myers: `medium.com/illumination/building-and-maintaining-trust-in-the-workplace-6da76ef1e369`

26

If You See Something, Say Something

"I have no right, by anything I do or say, to demean a human being in his own eyes. What matters is not what I think of him; it is what he thinks of himself. To undermine a man's self-respect is a sin."

—Antoine de Saint-Exupery

***

You've probably heard the phrase "If you see something, say something." It was developed by the Department of Homeland Security in 2010, and serves to encourage the public to protect our country. While the phrase was created to raise our awareness of the signs of terrorism and terrorist-related acts, it also serves to show the importance of reporting actions and behaviors that cause concern. When we use this motto in the workplace, we can help protect co-workers from sexist and racists acts, discrimination, and harassment.

There are many who, if asked, would say they are supporters and allies in the fight for equality. They will claim to not be racist, sexist, homophobic, or xenophobic and that it's impossible for them to express discriminatory beliefs because they don't have them. And while that mindset is a starting point, folks may mistakenly believe that because they are not intentionally perpetuating inequality in the workplace that they are in fact part of the solution. The problem with that position is that it is not active. It's just a mindset. Driving change requires action—actively addressing and challenging discriminatory acts and systems. That is why it is imperative that we be willing to "say something" when we see or hear something that denigrates others, even when we believe the person did not intend to do so. Saying something is doing something and an effective practice to educate people about the harm they are causing. I recall in the early years of my technology sales career, I overheard two White male colleagues having a discussion about me. They felt that the only reason that I was hired was because I was Black. During their chat, one said to the other, "Yeah, but she is pretty smart for a Black girl." I was stunned at their surprise that I could be smart, Black, and a girl, and appalled at their stance that I did not deserve to be there. Their comments were clearly sexist and racist, and my guess is that they had no idea. I had no inkling how to react, so I neither said nor did anything except strengthen my resolve to outperform them. I remember thinking that I should say something, but what? Whatever it was going to be, one thing was certain, it was going to come out wrong and more than likely lead to conflict. I remember feeling hurt and insulted that they had such low expectations of me and half-heartedly happy that at least one person on my team could see past my gender and race to recognize my intelligence. As I think back on how I would respond if this were to happen today, I would start by saying something like, "Excuse me. I could not help but overhear your conversation. I'm curious as to why you think that I was hired because of my skin color." I believe that calling his attention to what was said may start a dialogue where he could explore how he came to that conclusion. If we could get to the first base of exploration, instead of a dismissal that nothing was meant by it, I could then get to the second base of explaining how his remarks made me feel. If he has not shut down by that point, I could get to the third base of educating him on why his comments were harmful, and together we could finally take it home with an agreed upon path forward.

Everyone has been a hypocrite at some time or another, either consciously or unconsciously. During a meeting with several trusted White colleagues about philanthropic initiatives, we discussed sources for funding. One brought up a program from a prominent foundation that should be considered. Another colleague who was very familiar with the program responded that to qualify for the program, the business had to be minority owned, meaning that we would not be eligible. So instead of moving on, the colleague who first brought up the program joked that we say that Yvette owns the business—the only person of color in the meeting. A few folks laughed, and a few didn't. I was one of the ones who didn't laugh. The fact that she was comfortable saying the words was off-putting. However, the woman who made the comment was someone whom I respected and had worked with for years. I knew that she meant nothing by it but had taken note to call her about it later. Immediately following the meeting, another woman from the call reached out to check in on me as *she* was offended by it. She also asked whether I would be comfortable if she said something. My first inclination was to decline as I feel more than comfortable speaking up for myself. But I chose to accept her offer. I considered that the message may be better received from one White female colleague to another. The next day, the offender called with sincere apologies. We talked it through and became closer after that. Our relationship went from trusted colleagues to allies.

We can help ourselves and others by paying close attention to whether words and actions contradict our values and beliefs relating to equality. "Saying something" reinforces our own awareness and enables us to acknowledge as well as interrupt our own inappropriate behavior and that of others. When my colleague spoke up for me, she was in accord with her values and working to create a better workplace. When confronting discrimination of any type, we must be clear, direct, and confident in our approach to avoid ambiguity. Otherwise, we risk the message not getting through and not serving its intended purpose. The issues are too important to let a good opportunity go to waste. As you prepare to "say something," consider the recipient of your message. For some individuals, their actions seem out of character, and they may welcome the timely intervention. Others may believe themselves to be infallible, will admit to no wrongdoing, and perhaps will even blame you for being overly sensitive. A few seconds of forethought will help tailor a message in such a way that it can be received. The goal is not to change the individual, only make them aware. Change is an individual choice.

At times, speaking up can be difficult. A phenomenon known as the bystander effect hinders people from intervening in the moment when others are also witnessing acts of misconduct. A post in *Psychology Today* puts it this way, "Social psychologists Bibb Latané and John Darley attribute the bystander effect to two factors: diffusion of responsibility and social influence. The perceived diffusion of responsibility means that the more onlookers there are, the less personal responsibility individuals will feel to take action. Social influence means that individuals monitor the behavior of those around them to determine how to act." Today, the bystander effect is alive and well in the workplace, when people witness incidents of sexual harassment, inappropriate behavior, as well as intolerance or discrimination against women and individuals within the LGBTQ community and fail to speak out about it. When we choose to speak up, we're preventing someone else from experiencing discriminatory or abusive behavior in the future.

Michelle was caught off guard when a senior colleague told a sexist joke. The easy choice would have been to laugh along, go about her day, and not give it a second thought. After all, the joke came from someone in a managerial role. She thought that perhaps it would be harmless if she said nothing, but within minutes reconciled that the behavior and saying nothing could do plenty of harm. She reflected on recently updated HR policies, current media, and social movements like #MeToo, which crystalized the damaging effect this behavior can have on marginalized groups, and the workplace overall. Not taking a stand would be condoning the behavior and allowing it to perpetuate, not only by him, but others who find the behavior acceptable. Doing nothing would mean a missed opportunity to potentially prevent him from harming others. It would have been so much easier for her to craft a response if only the jokester had been a peer, but the fact that the individual was in a managerial role made her very uncomfortable. Yet she felt compelled to do or say something. By now it was way too late to act in the moment. She pondered whether she should speak to the jokester privately, request guidance from her manager, or go to human resources. She opted to go to human resources who thanked her for bringing the incident to their attention and assured her that the matter would be addressed. Afterward, she felt as if a huge burden had been lifted and her mind was now free to focus on work again. In the process of following her conscience, she became more alert to racist and sexist comments and now felt empowered that even she could make an impact just by listening and figuring out ways to respond within her comfort zone.

I've learned over the years that the best way to express a concern or challenge someone begins with how I start the conversation. How I start the conversation is a pretty good indicator of how it will end. When my approach is aggressive, judgmental, or tense, there's a strong chance that the individual will become defensive; we'll go back and forth with points and counterpoints and may never achieve a favorable outcome. My message will not be received unless I'm really lucky—which is hardly ever. It depends on the strength of the relationship. By contrast, when I approach calmly and with a collaborative mindset, a courageous conversation ensues, and I have a better chance of being heard.

Champions of inclusion understand that doing nothing is the absolute worst thing to do as we value the well-being of others. There is always something that can be done, and we look for ways to make impactful inclusion happen. Demonstrating an intolerance for acts of inequality may encourage others to do the same.

Actions

Translate Good Intent and Words into Meaningful Action

Many of us consider ourselves to be supporters of equality and would never intentionally discriminate or demean others who are different from us. Our actions speak louder than our words. Always be prepared to say something by doing the following:

- Having a few planned responses ready when you witness inappropriate behavior in the workplace to help colleagues realize their actions
- Understanding human resources policies and leveraging them when necessary
- Educate yourself on the bystander effect and its impact on individuals as well as the organization

Learn to Recognize Acts of Exclusion

It's easy to be unaware of acts of exclusion and their detrimental effects when they are not happening to us. We can't address them if we don't know what they are. Learning to recognize even subtle acts of discrimination or intolerance as in the stories depicted in this chapter will improve our ability to not only say something but help us know what to say. The more awareness we have, the more effective we can be.

Action Accelerators

- **HBR.org**: "How to Speak Up If You See Bias at Work," by Amber Lee Williams: hbr.org/2017/01/how-to-speak-up-if-you-see-bias-at-work

- `Catalyst.org`: "Interrupting Sexism at Work: What Drives Men to Respond Directly or Do Nothing?" by Negin Sattari, PhD, Emily Shaffer, PhD, Sarah DiMuccio, Dnika J. Travis, PhD: `www.catalyst .org/reports/interrupting-sexism-workplace-men`

- `FastCompany.com`: "4 women on finding the right response to sexism at work," by Gwen Moran: `www.fastcompany.com/ 90363669/4-women-on-finding-the-right-response-to-sexis m-at-work`

- `YouTube.com`: *Stand up, speak up*, Deloitte: `www.youtube.com/watch? v=DJvJAyZnAaA`

- `YouTube.com`: *Speak Up to Build An Equitable Workplace*, Crucial Learning: `www.youtube.com/watch?v=cTzqromPUVU`

Sources Cited

`PsychologyToday.com`. "Bystander Effect," `www.psychologytoday .com/us/basics/bystander-effect`

INTERVIEW: TELLING IT LIKE IT IS. . .TO GET WHERE WE NEED TO GO

Bystanders Do More Harm Than Good

Meet Delia. She is college educated and living her dream of working in the advertising industry. Jokingly, she considers herself a late bloomer as she delayed attending college until her late twenties. She began her career as an account coordinator and has proudly progressed to an account executive role after four years of hard work. Always willing to step outside her comfort zone in support of achieving goals has served her well. Now on a team of eleven creatives, she is one of two women on the team and is the only person of color. The demographics of the team are nothing new to her as the ad industry is plagued by a lack of diversity. She often finds herself battling bias, racism, and sexism in subtle ways from colleagues; her manager, Geoff; as well as clients. She laments how some days are worse than others and how internal systems are not designed to support victims but to protect the organization.

After 18 months of working to bring in her largest client to date, she was now in the negotiation stage. During the meeting with three cross-functional team members including Geoff, sexism slapped her in the face—almost literally. She had just completed presenting, and they were in the process of working out the details to close the deal. There were 12 people gathered around a U-shaped table. As Geoff was speaking, a male from the client's team leaned in to ask her a question directly. Before she could utter a word in response, Geoff stopped mid-sentence to order *her* to be quiet by yelling her name as if she was a child being scolded by a parent. In jaw-dropping shock and an onslaught of anger and humiliation, she froze. Geoff had to know that she was not the individual speaking. Certainly, after years of working together, he knew the sound of her voice. The fact that he literally yelled at her was bad enough, but to do it publicly, in front of the client, was the ultimate insult. She never felt more disrespected in all her years than she did in that moment. Did he not admonish the man because he was a man or because he was the client? The reason did not matter. She knew that the situation could have been handled better and that Geoff should have known better. Defending herself in the moment would have come across as unprofessional and perhaps would have made the situation worse in ways she had yet to imagine. The room went dead silent for a few seconds before Geoff continued. Delia was so busy trying to control her anger that she could no longer concentrate and engage at her best. For the remainder of the meeting, she found herself in pure disgust each time she heard Geoff's voice or looked his way. His attitude seemed cavalier. It was as if he reveled in his attempt to put her back in her place.

She reflected on how it would have made a world of difference that day if anyone in the room had acknowledged what happened by saying something as simple as, "Excuse me, she was not the one talking—that wasn't okay" or "Do you often speak to staff in that way?" Anything that would have caused him to pause, reflect, and apologize. She resented the member of the client team for not correcting Geoff and owning the fact that he was the one doing the talking that Geoff found so inappropriate. Ten bystanders observed the misconduct, many may have found Geoff's behavior unacceptable, and no one spoke up. Could it be that they were shocked as well and had no clue what to do? Perhaps her colleagues feared retaliation. Word of Geoff's behavior spread like wildfire back at the office. A few peers approached to take her side, privately of course. As far as she knew, Geoff had never been called to account for his actions that day and was free to repeat them. People suffer when no one stands up.

Avoid Common Terms That Divide Us

"Those who cannot remember the past are condemned to repeat it."

—George Santayana

***

Commonly used language in business has an inherent built-in bias toward masculinity and privilege, as business systems and structures were designed by and for the benefit of the majority culture. Our perceptions are shaped by what we see and hear, and the words we hear every day inform our reality. An article on HuffPost.com shares, "Language is very powerful. Language does not describe reality. Language creates the reality that it describes." Think about that. When we continually hear words and phrases, they consciously or unconsciously shape how we see things. Remnants of ideals from centuries ago remain in workplaces of today. Today, it's hard for many to fathom a time when women were considered property, not allowed to vote, or earn a paycheck. Though that time has passed, male chauvinism is alive and well in the workplace as women are objectified and receive less pay than men for equal work, signaling that women are less valuable. The marginalization of Black people was designed with intention with laws like Jim Crow, which legalized racial segregation and denied Black people the right to vote, hold jobs, and acquire an education, signifying that Black people are not worthy of equal rights based on skin color. Acts of anti-Blackness persist with the killings and beatings of unarmed Black men and women as individuals hold unfounded beliefs that Black people have inherently violent natures and should be feared. African American women are referred to as "angry Black women" for speaking their minds in the workplace. Black men are admonished by colleagues and bosses to be "less assertive" for displaying the same behavior as their White counterparts. Many hiring managers feel that they are "lowering the bar" when considering Black candidates. These beliefs are rooted in our history.

The Chinese Exclusion Act of 1882 banned Chinese immigrants from acquiring land, marrying non-Asians, testifying in legal proceedings, obtaining citizenship, and receiving legal protections guaranteed by the U.S. Constitution. Anti-Asian sentiments have persisted throughout the decades from American automobile makers unfairly blaming Japanese car manufacturers for their demise in the early 2000s to the breakout of severe acute respiratory syndrome (SARS) in 2003 and most recently covid-19 in 2020. Despite being characterized as the "model minority," people of Asian descent continue to experience hate crimes at an alarming rate as well as anti-Asian bias and discrimination in the workplace. Though they comprise 7 percent of the population in the United States according to a report by *The New York Times*, they remain underrepresented in senior leadership roles within American companies, media, and politics.

The oppression of Native Americans began when Christopher Columbus arrived in North America in 1492. A post on the site california-mexicocenter.org states,

---

"The British and French arrived to pillage the eastern coast of North America in the 16th century. The first permanent English [settlement] in Jamestown in 1607 was the beginning of their ambitions to establish colonies and take the land. Although most of the English brought families, they raped native women (regardless of the perpetuated mythology of the animated film *Pocahontas*) and began the practice of scalping. Some macho Europeans made a prosperous career by becoming 'Indian fighters' and scalping native children, women, and men. Settler militias plundered and burned native agricultural fields, destroyed to the ground native villages, and massacred at will. The term 'redskin' was coined to describe the remains of native people skinned alive and left to rot in the fields."

The oppressors believed that it was their right to oppress in the name of building for themselves a new world called the United States of America.

After decades of civil and human rights movements and changes in legislation, the demographics of the workforce have shifted to include women and other historically marginalized groups. However, offensive terminology created by men of days past persist. The workforce of today is plagued with derogatory language so commonplace that it's considered acceptable. No matter your industry, more than likely there is widely accepted jargon that has the potential to divide us. Consider *male/female* components, *blacklist/whitelist, segregation, blackout,* or *slave/master* in the tech industry. The common use of these terms perpetuates racism and sexism and thus is a contributing factor to the industry being less appealing to women and people of color. The technology industry struggles to attract and retain women, Latinx, and Black populations. Several tech giants have moved to replace such terms with *pin/socket, accept/reject, separation/division, restrict/downtime,* and *primary/secondary,* respectively. The real estate industry uses the terms *master bedroom* and *master bathroom* when describing the largest bedroom and private bath in a house. The word *master* has a strong association with slavery. As more of us awaken to the importance of inclusive language, many realtors are now replacing the word *master* with *primary.* According to an article on TheFederalist.com,

> "In addition to combatting 'racist' terminology, some real estate groups are pushing to phase out 'gender-specific' language such as 'man cave' or 'she shed' for the alternative 'accessory dwelling unit' or 'den' to avoid offending transgender or nonbinary people. Instead of 'mother-in-law suite,' realtors are supposed to say 'guesthouse' or 'in-law suite,' and any rooms that hint at the two sexes such as 'Jack-and-Jill bathroom' should be replaced with 'dual-entry bathroom.'"

Consider the practice of blind résumé reviews in human resources. While the point is to create more equitable hiring practices, the term can be offensive to individuals who navigate this world without eyesight. It's insensitive to compare viewing a résumé with blocked out information to the lived experience of a person who is blind. Similarly thoughtless are references to rambling conversations as "speaking Chinese," "opening the kimono" as a means of requesting that someone reveal information, gathering the team for a "pow-wow" as a means to call a quick meeting, or an individual with the least amount of authority as the "low man on the totem pole." While these terms may be as common as "conversations about the weather" for most people, they are insensitive and derogatory to African Americans, individuals with disabilities, Asian Americans, and Native Americans.

Let's not make the mistake of lulling ourselves into believing that words are being used without malice and no one cares about their hidden origins. As we work toward creating a more inclusive and racially aware society, altering language may seem like a superficial fix. When we get in to the habit of questioning whether the jargon of our industry includes or excludes, we are not making superficial gestures but foundational actions for building inclusion and belonging. The little things add up. Champions of inclusion accept responsibility for our words, consider their impact, and respond with humility when we make a mistake.

Actions

Develop an Anti-racist and Anti-sexist Vocabulary

Educate yourself and learn to recognize discriminatory, sexist, or racist terms used in everyday language as well as written communication. Make a list of new/replacement terms, update regularly, and share with friends and colleagues.

Change is constant, and new words and phrases crop up in mainstream media that we may be inclined to adopt without a second thought. In the spirit of inclusion, consider whether they may unintentionally cause harm before adopting them.

Resist the Temptation to Resist

The majority point of view is ingrained in systems, institutions, and individual beliefs. Embrace the fact that there is a minority point of view that is valid, holds weight, and is worthy of acceptance.

Anticipate Feelings of Awkwardness

Some of us may find ourselves within organizations with long histories of marginalizing members of underrepresented groups, and pivoting from old to new terms may feel like swimming upstream. Team members who are yet to start an inclusion journey may give us the side eye when we swap terms. It can feel awkward and compel us to stick with the status quo. Be prepared to explain new word choices, and invite others to do the same.

Action Accelerators

- **Ted.com:** *What It Takes to Be Racially Literate,* TEDWomen 2017: www.ted.com/talks/priya_vulchi_and_winona_guo_what_it_takes_to_be_racially_literate?referrer=playlist-talks_to_help_you_understand_r&autoplay=true

- **Today.com:** "These 11 everyday words and phrases have racist and offensive backgrounds," by Christopher Cicchiello: www.today.com/tmrw/everyday-words-phrases-racist-offensive-backgrounds-t187422

- **JDsupra.com:** "Racist Language and Origins I Didn't Always Know," by Gina Rubel: www.jdsupra.com/legalnews/racist-language-and-origins-i-didn-t-35616

- **FastCompany.com:** "The Problematic Origins of Common Business Jargon," by Lydia Dishman: www.fastcompany.com/90239290/six-business-phrases-that-have-racist-origins

Sources Cited

Dr. Frank Garcia Berumen. "The Most Marginalized and Impoverished People in the United States and in the Americas has Always Been and Continues to be Native Americans," California-mexicoCenter.org, July 1, 2020, www.california-mexicocenter.org/the-most-marginalized-and-impoverished-people-in-the-united-states-and-in-the-americas-has-always-been-and-continues-to-be-native-americans

Brittany Wong. "12 Common Words and Phrases with Racist Origins or Connotations," HuffPost.com, Updated July 8, 2020, www.huffpost.com/entry/common-words-phrases-racist-origins-connotations_l_5efcfb63c5b6ca9709188c83

Denise Lu, Jon Huang, Ashwin Seshagiri, Haeyoun Park, and Troy Griggs. "Faces of Power: 80 percent Are White, Even as U.S.

Becomes More Diverse," `NyTimes.com`, September 9, 2020, `www.nytimes.com/interactive/2020/09/09/us/powerful-people-race-us.html`

Jordan Boyd. "You Can't Say 'Master' Bedroom Anymore Because Some Realtors Decided It's Racist," `TheFederalist.com`, August 10, 2021, `thefederalist.com/2021/08/10/you-cant-say-master-bedroom-anymore-because-some-realtors-decided-its-racist`

Reframe Difficult Conversations on Polarizing Topics

"You must let suffering speak if you want to hear the truth."

—Cornell West

***

Social inequity is more pronounced now than in previous decades—or at least more of us have become aware of it. We never know what's going to be served in the workplace from day to day. Between acts of hate and violence and movements to maintain or disrupt power, we can be just minutes away from crises or chaos at any moment in time. Sure, the work is constant, but the conditions under which we must perform can vary greatly from person to person. When a casual conversation with a colleague suddenly shifts to race, a polarizing political view, legislation impacting the LGBTQ+ community, or something else that makes us highly uncomfortable to talk about, we tend to avoid it all costs as such discussions can become extremely emotional (see Figure 28.1). We fear that anger and resentment may build or we'll be misunderstood, say the wrong thing, be thought of as racist, sexist, or homophobic, or worse—lose our job. Work relationships may be forever damaged. When the fight or flight response kicks in during difficult conversations, many will choose the flight every time. Rather than muster the courage to lean in, the option to quickly change the subject or abruptly remember that we're running late for a meeting lies just beneath the surface and is far more appealing. After all, we are wired to choose pleasure over pain.

Figure 28.1: Courageous conversations trigger emotions.

The degree to which we welcome or avoid conversations that we find difficult determines whether we are building or blocking a culture of inclusion. Before running from the next opportunity to have a difficult conversation, consider the impact of *not* having it—a culture of silence that festers and diminishes one's sense of belonging, reduced productivity, and the potential to negatively impact one's mental health over time. Reframing these interactions from difficult conversations to courageous conversations is more empowering. It helps us to boldly be part of the solution. This is our opportunity to do better. Courageous conversations involve intentionally giving space to complex issues. They are courageous in that they require bravery to sit in discomfort long enough to explore one another's feelings and vulnerabilities—especially when racial and social injustice are prevalent or dominating the media. The challenge of doing the work to uproot assumptions, not judge, share our experiences, and hear the experiences of others is part of the process. Accept it. Courage and self-awareness are required if we are ever going to connect, understand another's lived experience, and be understood. Courageous conversations can bring about allyship, increased awareness of the impact of a situation, and validation of difficult experiences. Yup. That mass shooting did just happen, killing innocent people of color. Yup. It's the second one in as many months. Yup. Nothing yet has brought about change. Yup. The next time it may be me or a loved one. Yup. These feelings of fear, helplessness, anger, and pain have followed me and/ or a co-worker into the workplace. Yup. Here, many of us sit trying to simply get through the day under a cloud of emotions and are expected to produce as if nothing has happened. Considering that this narrative may be the dialogue playing in one's mind, the fact remains that we must be open to other people's experiences that are not necessarily similar to our own. We must be mindful of the multiple perspectives born from an individual's background, experience, and identity. While you may be a White person who has never been profiled and threatened by a police officer on your way to work, the idea that that just happened to your Black colleague simply because they were Black may be something that is completely alien. White people may now feel villainized for the first time and unsure of how to interact with people of color as awareness of the atrocities marginalized groups experience at the hands of Whites now and throughout history is more prevalent. These feelings and attitudes have increasingly become a part of workplace culture, and having courageous conversations is the best way to hear and be heard so together we can figure out a way forward (see Figure 28.2).

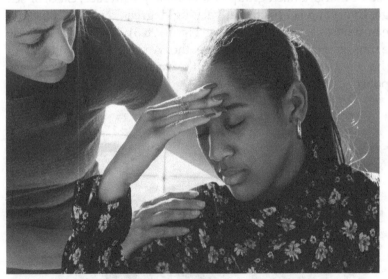

Figure 28.2: Lack of courageous conversations diminishes productivity.

Sadly, many in the workplace remain oblivious to the aftermath of hate-based violence, laws that further marginalize underrepresented groups, and the massive civil unrest as it does not affect them. Walking away or putting our heads in the sand is not a privilege that everyone has. We must approach with a curious mindset and ask ourselves soul-searching questions like these:

- How would I feel if I were in this situation?
- Why do I care, and what do I care most about?
- How have I been personally impacted?
- What are the roadblocks to change within the organization that I can influence? (Nothing is too small.)
- What am I willing and able to do to empower/support/advocate for others?
- How can I leverage my power and privilege to improve someone's experience today?
- What do I lose when I avoid courageous conversations, and how could my life improve when I embrace them?

The most valuable work happens between people who are willing to understand and learn from one another. It is possible for each of us to support one another during social unrest whether we're from an oppressed or marginalized group or a visionary with race, class, or sex privilege. Everyone is affected at varying degrees regardless of who they are or where they come from (see Figure 28.3). The impact of these divisive times is a lot like the air we breathe. We all are in it, and we can choose how we want to respond with the awareness that no response is a response.

Figure 28.3: Remember remote colleagues.

Actions

Become Skilled at Having Courageous Conversations

Courageous conversations are unchartered territory for most of us. Knowing how to have them will lead to better outcomes. Access tools to help shape and guide conversation. Check out the following resources:

- **The Denver Foundation:** www.nonprofitinclusiveness.org/ agreements-courageous-conversations-and-active-learning

- **PsychologistsOffTheClock.com:** Podcast episode 96. *Effective Conversations About Diversity* with Anatasia Kim and Alicia del Prado, by Debbie Sorensen, offtheclockpsych.com/diversity- conversations/#:~:text=Discussions%20of%20diversity%20 issues%20are%20more%20critical%20than,leading%20to%20 anger%20and%20hurt%20on%20both%20sides

- **Blog.XYplanningNetwork.com:** "How to Have Difficult Conversations About Diversity," blog.xyplanningnetwork.com/ advisor-blog/how-to-have-difficult-conversations-about- diversity

Leverage Your Self-Awareness Developed in Activity 1 and Activity 2

When we are self-aware and in touch with our *why*, it is easier to have courageous conversations. Understanding our emotions helps us to view situations more objectively and avoid responding in ways that may be perceived as defensive. We must be open to the possibility that the way in which we see things is not necessarily the right way or the only way.

Say the Unsaid

Resist the posture of walking on eggshells when navigating these conversations. No tiptoeing around the issues. The proverbial "elephant in the room" will remain until acknowledged. When we acknowledge and accept the fact that the conversation can trigger feelings of discomfort and strong emotions, we are not surprised when they do. Thus, we are better prepared to relive it. Bring empathy to the conversation by considering the other person's frame of mind. Remember to breathe to manage the anxiety.

Create Space to Listen to One Another

Bear in mind that everyone is the expert on their own experience. Trust that what they are sharing is valid. Avoid judging and blaming, as this may be pointing out fallacies in our own thinking. Our role is not to alter their perspective or prove whether something is true. It is true for them, and that is what matters most. Honor the space with humility, trust, and respect.

Action Accelerators

- **Medium.com:** "How to Talk About Racism at Work, Hint: It's Not as Scary as You Think," by Ajah Hales: medium.com/our-human-family/how-to-talk-about-racism-at-work-7f27da56158a

- **HBR.org:** "High-Performing Teams Need Psychological Safety. Here's How to Create It," by Laura Delizonna: hbr.org/2017/08/high-performing-teams-need-psychological-safety-heres-how-to-create-it

- **Book:** *We Can't Talk about That at Work! How to Talk about Race, Religion, Politics, and Other Polarizing Topics*, by Mary-Frances Winters: www.amazon.com/Cant-Talk-about-That-Work/dp/1523094265/ref=sr_1_3?crid=3JBCCP4867TW1&keywords=mary+frances+winters&qid=1653145706&s=books&sprefix=mary+frances+winters,stripbooks,122&sr=1-3

- **YouTube.com:** *What White people can do to move race conversations forward*, Caprice Hollins, TEDxSeattle: www.youtube.com/watch?v=7iknxhxEnlo

29

Embrace Gender Identities

"Gender identity belongs to the person who lives it, but one cannot deny that observers will make assumptions about us based on their understanding or comprehension of gender signals."

—Jamison Green

***

Growing up in the 1960s, '70s, and '80s, I was aware of only two genders—male and female. When I entered corporate America in the mid-'80s, I came to know colleagues who identified as gay or queer. However, male and female remained their gender identity as far as I knew. When I could not discern gender, I learned that individuals described themselves as androgenous and that it is completely inappropriate to satisfy our curiosity by asking them about their gender. As I socialized in new circles and traveled the world, I was slowly awakened to my narrow perspective of gender. Continual focus through that lens is highly exclusive of the beauty of individuals that I've had the pleasure of meeting and who identify as *genderfluid*, moving between gender identities; *agender*, having no specific gender identity; and *bigender*, having two gender identities. Gender identity is a personal sense of self. It's how someone conceptualizes their own gender. Getting familiar with language and terminology is a sign of respect and positions us to better understand and support our LGBTQ colleagues.

If you grew up like me with an awareness of two genders, becoming gender fluent is going be confusing at first. My children came out almost a decade ago, and I am still learning because I want to. They deserve the best mom I can be. Our co-workers deserve our sincere effort and an open mind as we all work together toward equity and acceptance. Words and their definitions change or become refined over time. The acronym LGBTQIA+ (Lesbian, Gay, Bisexual, Transgender, Queer/questioning, Intersex, A-gender, plus other identities) continues to evolve and is an inclusive term encompassing individuals of all genders and sexualities.

As we continue to build our inclusive mindset, the introduction of new identities and learning their nuances will come with discomfort and fear of saying something wrong. We'll make lots of mistakes and learn from them before confidence takes over. The words we choose represent what we believe and drive our interactions with one another. Impactful inclusion is about creating experiences where everyone can be their authentic self and feel valued and respected for the differences they bring. Those with strong religious beliefs opposing the LGBTQIA+ community must be mindful that beliefs should not be asserted in the workplace as they can diminish one's sense of inclusion and belonging as well as potentially damage working relationships. When we do not actively include, we exclude. We play a big part in making it easier for co-workers to feel a sense of belonging. Clarifying our gender pronouns in the workplace is one way we can demonstrate awareness and respect while helping to deconstruct the gender binary. Those she/her/hers, he/him/his, and they/them/their in email signatures, social media profiles, and introductions of ourselves may seem odd, uncomfortable, and grammatically incorrect in some cases when we speak them aloud and in written communications. But those signals are extraordinarily powerful to those who are gender diverse. If you have never had to concern yourself with pronouns used when others refer to you, that is a privilege that not everybody gets to enjoy. When people are referred to with the wrong pronoun, they can feel disrespected, invalidated, and alienated.

When I was seven years of age, I was infatuated with the boy next door. Well, actually the man next door. He was 19 and the most gorgeous human I'd ever seen. We bonded instantly once we realized that we shared the same birthday. I loved spending time in his apartment, which was directly across the hall. He lived with his White mom and

Black dad, and they treated me like family. They gave me the kind of snacks that I could never get at home, played checkers and tic-tac-toe with me, talked with me instead of at me, and in hindsight, I felt like they got me as the little lonely kid that I was.

What I loved about Paris (not his real name) was that he was unlike anyone I'd ever met. He always greeted me with the biggest smile and made me feel so special. The world stopped whenever I looked at him. I remember how he always smelled so good, and his style of dress was so cool to me. He wore earrings—sometimes small hoops or gold studs, colorful headbands on his locks of curly hair, brightly colored shirts, and bell bottom jeans, which laid perfectly over his sandals or sneakers. His walk was more like a prance. I was truly in awe and found myself watching for him through our front window just to get a glimpse of his awesomeness as he entered and exited the building. He was definitely the man I would marry when I grew up. A few weeks went by where I didn't see him, and there were a lot of unfamiliar people coming and going from our building. One afternoon, I overheard my grandmother consoling Paris's mother through a broken heart and heavy sobbing. His mom moaned through her tears that he had hung himself and she'll never understand why he loved another man. I did not know how to process what I was hearing. I had no clue what it meant to "hang yourself" nor did I understand the confusion she had about loving another man. Isn't loving someone something that everybody does? I continued to listen just outside the door as two friends were trying to make sense of something so senseless. No one ever told me that I would never see Paris again or even noticed that I was mourning the loss of him too. As I got older, I never forgot him. His spirit was so vibrant. As an adult, my life's journey has introduced me to new Parises, and I have experienced his beauty through them. In their stories I feel certain that Paris was more than likely emotionally tormented as he struggled for acceptance and endured harassment. Back in the 1960s, I can only imagine how especially difficult life must have been to be both Black and gay, and I believe that he ended his life because of it. While the struggle for equality and belonging is still real, I can't help but think that Paris might still be shining his light in this world if change had come sooner. Perhaps we'd be the best of friends, and he could have hung out with me and my children. Maybe something as simple as using a pronoun in support of his gender expression could have helped him feel more accepted, less tormented, and more willing to embrace life.

A post from FightHatred.com shared,

"In general, however, a little over half of Americans, or 52 percent, claim to be comfortable with using gender neutral pronouns when referring to a person who requests its usage, while 47 percent of Americans say that this makes them uncomfortable. In fact, 6 out of 10, or roughly 61 percent of Americans aged 18 to 29 report feeling very comfortable using gender-neutral or non-binary pronouns, a sharp contrast to 48 percent of Americans aged 50 and older who report the same."

Create space for gender identity by asking co-workers, especially when they're new hires, their pronoun. It's as simple as, "Hello, I'm Tina, and my pronouns are she/her. Happy to meet you. What are the pronouns you go by?" We should never assume someone's gender identity while being aware that a person's pronouns can change over time. To ease into it, use "they" until you learn more. It's important to recognize that gender is not binary, and not everyone's gender conforms to the one they were assigned at birth.

Actions

Demonstrate Acceptance and Support

Create social norms around pronoun use. Add your pronouns to your email signature, social media profiles, professional bio, business card, and name tag at conferences or events.

Where it makes sense to do so, include your pronouns when you introduce yourself.

Recognize Impact Over Intent

Assuming someone's gender and getting it wrong can be upsetting or offensive. The defensive response of "not intending harm" or "being too sensitive" cannot and does not erase the impact. We must accept responsibility for the mistake, genuinely apologize, learn from it, and continuously work to improve.

Over the next several weeks, take note of how many times you refer to someone without using their name. Consider whether you were accurate in the use of their pronouns. Ask yourself, how often did I assume?

Celebrate International Pronouns Day

Mark your calendar to celebrate International Pronouns Day on the third Wednesday of October. Visit Pronounsday.org for details and resources.

Be a Continuous Learner

According to Healthline.com, "There are so many gender terms out there, many of which overlap. Some also have definitions that shift over time or across different sources of information." That said, check out the MSN post "68 Terms That Describe Gender Identity and Expression," by Mere Abrams, LCSW, and Sian Ferguson, at www.healthline.com/health/different-genders.

Action Accelerators

- **MyPronouns.org**: "Asking Others Their Pronouns": www.mypronouns.org/asking

- **LaborEmploymentReport.com**: "Pronouns and Coworkers and Misgendering (Oh My!)," by Elizabeth Torphy-Donzella: www.laboremploymentreport.com/2020/01/29/pronouns-and-coworkers-and-misgendering-oh-my

- **YouTube**: *How to Share Gender Pronouns at Work*, Culture Amp: www.youtube.com/watch?v=trmzT-N7XnM

- **YouTube.com**: *Why Gender Pronouns Matter*, Mala Matacin, TEDxHartford: www.youtube.com/watch?v=hsHXyGiCk6g&t=48s

- **Workology.com**: (Podcast) Episode 169, "Inclusion and the Use of Gender Neutral Pronouns in the Workplace," by Jessica Miller-Merrell: workology.com/ep-169-inclusion-and-the-use-of-gender-neutral-pronouns-in-the-workplace

- **JenniferBrownSpeaks.com**: "The Performance of Gender: Improv and the Exploration of Identities and Norms with Ali Hannon": www.stitcher.com/show/the-will-to-change-uncovering-true-stories-of-diversity-and/episode/e146-the-performance-of-gender-improv-and-the-exploration-of-identities-and-norms-w-ali-hannon-81196486

Sources Cited

FightHatred.com. Gender Neutrality: What it Is and How to Be More Like It, October 30, 2019, `www.fighthatred.com/lgbt-rights/` `gender-neutrality-what-it-is-and-how-to-be-more-` `like-it`

Make the Workplace Safe for Everyone to Be Themselves

"It ain't enough to talk of toleration! Each of us is to be the vessel of unification."

—**Abhijit Naskar**

***

Increasingly, members of marginalized groups are aspiring to bring their whole selves to the workplace. The workplace becomes more enjoyable, and we can be much more productive when we can be our authentic selves. We wouldn't have to put energy that could be better served elsewhere into being someone who we are not for 40+ hours a week. Historically, individuals from underrepresented groups were forced to assimilate to the dominant culture simply to be employable, and things haven't changed much over the years. African American people have altered their natural hairstyles, immigrants have adopted American-sounding names, gay people remain closeted, and mixed-race individuals who can pass for White may choose to do so at the expense of embracing the richness of other cultures that make them who they are. Individuals with disabilities choose not to disclose for fear of being stigmatized. Those who are traditionally excluded have endured laughing at inappropriate jokes, laughing at jokes without a clue as to why they were funny, or dumbed ourselves down to not outshine White bosses. When trauma strikes through acts of violence on the innocent, the traumatized are expected to leave emotions out of the workplace. Shedding parts of ourselves at the door that have nothing to do with job performance may mean leaving the best parts of ourselves elsewhere. I can't begin to tell you the number of times I kept my perspective to myself when I found flaws in ideas or strategies from leaders and chose the safety of not speaking up for fear of retaliation; or I struggled to suppress emotions in front of colleagues when innocent people were senselessly murdered. Grace is sometimes needed to be our authentic selves. Many of us withhold a big part of who we are as individuals and resist being vulnerable to be seen as professional, competent, successful, and likeable.

So often, members of marginalized groups bring another version of themselves to the workplace: the work persona that makes us more acceptable. But what if the workplace was an environment where we could reveal more of ourselves? What could be possible that was once impossible? What would the impact of that be? True inclusion is when we are not compelled to conceal any aspect of who we are. When connecting on a personal or professional level, we would not need to censor or guard ourselves. However, showing up in our true form to a workplace that isn't conducive to receiving it may mean putting oneself at risk. For some, being open about their religion or sexuality can mean a daily struggle in a hostile work environment, losing their job, being bullied, or being passed over for promotion. It's impossible to present the real version of ourselves when our survival depends on withholding it. To be clear, showing up in our true version does not mean being unapologetic for exclusive behaviors, attitudes, and beliefs even though they represent one's true self, nor does it mean that we can bring the parts of ourselves that invite chaos and conflict. When we bring our authentic selves to the workplace, we freely bring emotions, thoughts, and ideas that may be counter to the dominant thinking without fear of expressing them. We can bring our unique perspectives and experiences to problems and ask questions or make mistakes without fear of judgment. This fear factor diminishes the quality and connection of working relationships, and as a result, the organization misses out on valuable contributions.

The ability to simultaneously be yourself and be accepted in the workplace is a privilege that is enjoyed by members of the majority culture who are White, cisgendered, heterosexual, and able-bodied males. When the environment is a reflection of who you are, it's much easier to be yourself. When we don't fit neatly into the mainstream, we must focus our energy on bridging the gap between who we are and who the majority culture expects us to be, which is more like them, and that's not achievable.

For Gina, a White, transgender engineer, bringing her whole self to work means no longer suppressing her real identity and being just as respected, valued, and accepted as a woman as when she presented as a man. She now struggles for the first time with being heard and taken seriously by most—a common experience of women in general. The support she gets from female colleagues and a few male allies is tremendous, and they make her feel like family.

For Robert, a coder with autism and father of two, bringing his whole self to work means getting smiles and giggles from colleagues when his children pop in on video team calls. Parenting is a proud part of who he is, and he enjoys allowing co-workers to meet his kids and giving his kids the fun of waving "hi" to the heads in the boxes on-screen. He feels seen beyond being an individual with autism and is comfortable bringing the parent part of himself to work.

For Hassan, a millennial graphic designer of Islamic heritage, being his authentic self means satisfying his natural curiosity with technology and contributing to projects on cross-functional teams where his ideas are explored and not discounted because of his age, race, or culture. The more he expresses his technical side, the more he contributes. He feels like his co-workers "get him" when they seek his opinion on tech stuff or show off a new tech toy. His love of tech is a part of who he is, and he is happy that he can unleash it in the workplace.

For Freida, a bookkeeper who is African American, bringing her authentic self to work meant working up the courage to stop covering her natural hair with straight hair wigs at the office. The straight look did not reflect her free spirit, and it felt unnatural having to wear it. She knew that in revealing her natural hair, her non-Black co-workers would react weirdly, and the boss may deem it unprofessional. The first week was hard. There were stares, intrusive questions, and folks wanting to touch. After a few weeks, her hair was no longer a topic of conversation, and she could get back to focusing more on work. She could now show up in a way that reflected who she was, and it felt liberating.

There are organizations whose diversity, equity, and inclusion efforts are perfunctory. They only give lip service to inclusion with symbolic gestures like cultural holiday celebrations, or they publicly commit to actions that are never taken or demonstrate support for social causes only on social media, while no one in leadership walks the talk. Environments like this are a long way from providing the conditions that make it safe for people to bring their whole selves to work. Champions of inclusion have an opportunity to create cultures of kindness that invite the whole person in. When we look at the whole of an individual—what makes them who they are—and respect all their dimensions, we are letting them know that they don't have to shed part of themselves in our presence. I'm not saying that we have to like all dimensions, but we must always show respect and kindness. Our individual and collective actions can create cultures rooted in respect, kindness, and authenticity, which is the key to being able to bring our whole selves to work.

Actions

Get to Know What You Don't Know

In thinking about your own organization or team, consider dynamics that may be blocking the psychological safety of others and preventing them from bringing their full selves to work. Reflect on the stories of Gina, Robert, Hassan, and Freida in this chapter as you explore the forces at play. Do you see co-workers reflected in their stories? As an individual, ask yourself, "What can I do to create a safe environment for co-workers, and what should I stop doing?" The responses identify gaps and create the foundation from which to begin building.

Create Safe Havens to Connect

Consider fun ways to get to know colleagues on a more personal level. The more ways we get to know them, the better the connection and opportunity to establish trust, allowing us to share aspects of ourselves that may not come up in day-to-day interactions. Create a quirky titled Slack channel that invites co-workers to share what they did over the weekend, swap recipes, post vacation photos, and do other activities that represent more aspects of themselves. Make sure to engage and keep the conversation going. This is an awesome way to bring in remote colleagues.

Create online watercooler huddles or coffee breaks via video conference. Invite the contacts from other departments or team members for a regular 20-minute get-together with no agenda and with only one rule: to not talk about work stuff or complain. Get the conversation started with icebreaker-type questions, for example, "As a kid what did you want to be when you grew up?" Let the conversation flow.

Leverage the safe space and collective wisdom inherent in organizational affinity groups also known as *employee resource groups* (ERG) and *business resource groups* (BRG). These groups are comprised of individuals with common interests and their allies. Raise topics and ask questions to begin a dialogue to address what's needed for everyone to be who they are in the workplace.

Action Accelerators

- **Workplace.msu.edu:** "Psychological Safety and DEI": workplace.msu.edu/psychological-safety-and-dei

- **YouTube.com:** *Why Workplace Culture Is Everyone's Responsibility*, SHRM: www.youtube.com/watch?v=P95eeEPCmlQ

- **YouTube.com:** *3 ways to create a work culture that brings out the best in employees*, Chris White, TEDxAtlanta: www.youtube.com/watch?v=2y8SA6cLUys&t=603s

- **JenniferBrownSpeaks.com:** *When Silence isn't Golden: Being Brave, Not Perfect, with Space and Grace*: www.stitcher.com/show/the-will-to-change-uncovering-true-stories-of-diversity-and/episode/e108-when-silence-isnt-golden-being-brave-not-perfect-with-space-and-grace-74426312

INTERVIEW: TELLING IT LIKE IT IS. . .TO GET WHERE WE NEED TO GO

People Don't Leave Companies, They Leave Cultures

Belinda is an accountant with 26 years of experience and has built a repertoire of knowledge and skills in various industries. Now employed by a small transportation company, she is the only African American woman in the department, which is led by two White males and a Latino male, Javier, to whom she reports. While no manager is ever all she needs them to be, she found that Javier's leadership style left a lot to be desired. From her perspective, he was incompetent and seemingly protected by the old boys' network. His ineptness appeared to go unnoticed by leadership but was very apparent to all direct reports. When he made mistakes, he never accepted responsibility and placed the blame on the shoulders of others—including hers. One thing for sure about Belinda is that she is not the type to sit silently when treated unfairly. Her collaborative approach of "It's not me against you but us against the problem" usually yielded a favorable outcome. Whenever she approached him to discuss her concerns, he was always dismissive with responses like, "Don't worry about it, it's no big deal" or "Let's talk about it later," and later never came. Productivity started to suffer as she found herself double- and triple-checking her work to make sure that it was accurate. Anyone can make mistakes, and though she worked hard to avoid them, she accepted responsibility and held co-workers to the same standard of "Know your stuff, and do your best." She believed that individuals in general, but especially those in managerial roles, should not be permitted to constantly drop the ball, pass the buck, and remain employed. By contrast, it was in her nature to exceed expectations. It was stressful and frustrating for her to work without feedback or the opportunity to get clarity on how she was to blame for his errors. How would she ever improve? She often wondered how she was ever to add value to the team as well as to the organization in that environment? Her attempts to provide input and contribute were consistently disregarded. There was a constant struggle to stay motivated without the support needed to perform to the best of her ability. Belinda's idea of bringing her whole self to work was feeling accomplished for a job well done, asking for help without judgment, and working through blockers to performance, none of which she was getting.

In one instance, and we'll call this one the straw that broke the camel's back, she was assigned to reconcile a project from her predecessor. It was purely busy work from her point of view. Regardless, she put her head down and got it done based on the instructions given her by Javier, instructions that she would soon discover to be flawed. When the result didn't meet expectations, Javier exploded at her desk and in front of colleagues. Anger was an understatement, and humiliation was just the tip of the iceberg. She froze in silence. His antics were especially hurtful as Javier was young enough to be her son. To maintain her composure and not respond in kind, she excused herself from her desk to calm down in the restroom. He was not at her desk when she returned. In fact, weeks went by before he faced her. They communicated strictly by email. She grew to prefer this way of interaction as it created a paper trail and allowed her to keep a safe distance. Their relationship continued to become more strained over time, and she was tired of fighting an uphill battle. The small amount of respect she held for Javier before that time vanished the moment that he berated her. She also lost respect for the entire leadership team as Javier seemed untouchable. She considered in jest that perhaps he held something incriminating over their heads. Why else would he be allowed to stay? Realizing that things were never going to change, she began her search for a new job. She survived the rest of her tenure buoyed by the thought that her misery would soon come to an end.

Don't Believe Your Lying Eyes—Test Assumptions

"The eye sees only what the mind is prepared to comprehend."

—**Robertson Davies**

***

Every culture has career advice passed through the generations. In the Black community, we were taught that we had to know twice as much to go half as far. It was a clear message that we need to outperform White people in every regard to have a chance at a fraction of their success. I've heard from my LatinX friends that they were advised as well to keep their head down and work harder than everyone else. My Asian friends felt pressured from an early age to aspire to attend top-tier universities as a means to acquire wealth and success as there was no other way. There's clearly a pattern of belief that people of color are disadvantaged from the start and must somehow overcompensate to have a chance at success. It certainly helps to have these pearls of wisdom from our loving parents who only want the best for us. Without it, who knows where we would be? So, many of us did exactly what we were told, and, in some respects for a small minority, it paid off as a component of an overall strategy. However, the majority of minorities in corporate America continue to face insurmountable barriers, despite the advice. Research from Statista.com shares, "Out of 500 current Fortune 500 CEOs, just five are Black while 466 are White." They continue with the facts:

> **According to data collected by *The Wall Street Journal*, just 1 percent of all CEOs of Fortune 500 companies are Black—a statistic that has been trending down over the last decade. From 2000 to 2010, the percentage of Black CEOs of Fortune 500 companies doubled from 0.6 percent in 2000 to 1.2 percent in 2010 before falling back to 1 percent in 2020.**

Asian-American CEOs have risen slowly over the last two decades, going from 1.8 percent in 2000 to 2.4 percent in 2020. Hispanic CEOs have seen a greater rise since 2000, going from 1 percent to 3.4 percent in 2020.

The rise in female CEOs has been perhaps the most significant over the last two decades. With just two women as CEOs of Fortune 500 companies in 2000, there are now 37 women in CEO positions as of August 2020—a leap of over 18 times.

Studies cited by *The Wall Street Journal* show how Black, Asian, and Hispanic executives face more obstacles when navigating the leadership pipeline of various Fortune 500 companies. Overall, surveys show that minority employees at these companies see obstacles like fewer opportunities for advancement, smaller rates of retention, and tougher recruitment barriers.

When we look at general diversity in the labor market as a whole, Zippia.com reports, "When broken down by race and ethnicity, Whites make up the majority of the workforce." They continue:

> **Federal studies have determined that the nation's labor force is 78 percent White, 13 percent Black, 6 percent Asian, 2 percent Biracial, and 1 percent American Indian. Additionally, people of Hispanic or Latino ethnicity, who may be of any race, represented 17 percent of the workforce.**

Women make up more than half of the United States' labor force. According to the U.S. Census, women represent 58.3 Percent of the U.S. workforce, while men represent 41.7 Percent.

Considering these statistics there's no wonder that the default assumption is that if you are a member of historically marginalized groups, that you have achieved less success than White counterparts. Couple this with unconscious bias, and you'll find that for those minorities that have broken through the status quo are on the receiving end of unconscious demotions—the habit of assuming that someone holds a lower position or career status than they actually do based on race, ethnicity, gender, or ability.

Consider stepping into a business meeting, turning to the youngest looking woman in the room, and requesting that she fetch your coffee. When she declines, explains that she is the newly hired director of engineering, and recommends that you fetch your own coffee, you are probably embarrassed as a result. But consider how she must feel. She was instantly demoted to a lower position purely on the fact that she is a woman with a youthful appearance, perhaps mistaken for an intern. Saying nothing continues to perpetuate and condone inequality. Over time, these consistent snap judgments can have a negative effect on one's psyche and signal that they don't belong in the career they occupy. When this happens once in a while, it's easier to dismiss, but having to correct people time and again can be exhausting. This instance is more commonplace than one may think, and it can happen to any member of historically marginalized groups even at the most senior levels. A post from Madamenoire.com shares, "Black women who have risen through the professional ranks are often mistaken for assistants or people in much lower positions, and time has shown no one, no matter their level of achievement, is immune." Successful investment firm executive Mellody Hobson often tells of an incident that happened about a decade ago when then-aspiring senator Harold Ford asked her to make some media connections for him. When she and Ford showed up at the offices of a major NYC media company, they announced to the receptionist that they were there for the lunch, a lunch Hobson had organized. The receptionist mistook them for servicers hired for the lunch and asked them where their uniforms were. The post goes on to say, "In my research, I've found that Black women are unconsciously demoted to even lower positions than White women," explained Dr. Suzanne Wertheim, researcher and founder of Worthwhile Research & Consulting. "For example, while White female doctors were mistaken for nurses, a Black female doctor was mistaken for an orderly. And while White female lawyers were mistaken for legal secretaries, Black female lawyers were mistaken for criminal defendants."

Interrupting these patterns means creating new cognitive habits and challenging assumptions about others. Rather than assume that an individual is part of the wait staff, an intern, or janitor, simply ask in a polite nonjudgmental tone, "So what do you do here?" or introduce yourself first, and then pose the question. Champions of inclusion rethink images of what accomplishment looks like, purposefully test assumptions, and step in as a buffer to correct the error when it happens in our presence. I am of the mindset "Better safe than sorry." Why not elevate the status of another? An individual just may become inspired to rise to the elevation. Imagine if we did that more often!

Actions

Be Aware of Occupational Stereotypes

In male-dominated industries or roles, women are easily demoted as the default is to think of a man. Women and people of color who are surgeons are assumed to be nurses or orderlies, attorneys are assumed to be paralegals or legal secretaries, CEOs are assumed to be administrative assistants, and school principals or superintendents are assumed to be teachers, teacher's aides, or janitors.

Break Automatic Associations

When seeing or meeting someone for the first time, bias and stereotypical beliefs can drive the interaction and cause us to unconsciously demote people. Invest the time in identifying members of marginalized groups who are high achievers to gain new context and create new thought pattens. Consider Eric Yuan, inventor of Zoom conferencing and member of the AAPI community; Dr. Tarika Barrett, CEO of Girls Who Code and member of the African American community; Satya Nadella, CEO of Microsoft and member of the Indian American community; or Beth Stevens, PhD, neuroscientist, MacArthur Fellow, and is White. Leaders and rising leaders are all around us, we just need to slow down and look.

Don't Credential Check

When we discover that an individual has a higher career status than we expect, often our surprise will prompt the need to verify their competence. Doing so signals that they must prove themselves, but why should they have to? Avoid intrusive questions about their background, experience, expertise, and education. Just go with the flow or check their LinkedIn profile. Never expect someone to validate their competence.

Action Accelerators

- **Monster.ca**: "How to Reduce Stereotyping in The Workplace," by Mark Swartz: www.monster.ca/career-advice/article/overcoming-workplace-stereotyping

- **FastCompany.com**: "The Common Habit that Undermines Company Diversity Efforts," by Dr. Suzanne Wertheim: www.fastcompany.com/3060336/the-common-habit-that-undermines-organizations-diversity-efforts

- **YouTube.com**: *Implicit Bias - How We Hold Women Back*, Maureen Fitzgerald, TEDxSFU: www.youtube.com/watch?v=YM1MBSczyzI&t=868s

- **YouTube.com**: *It's About Time We Challenge Our Unconscious Biases*, Juliette Powell, TEDxStLouisWomen: www.youtube.com/watch?v=thkmVv54e6M

- **YouTube.com**: *3 Design Principles to Help us Overcome Everyday Bias*, Thaniya Keereepart, TEDxPortland: www.youtube.com/watch?v=K6dstCUWsFY&t=104s

Sources Cited

William Rogers. "Little Progress for Black CEOs in the U.S.," Statista.com, September 30, 2020, www.statista.com/chart/23060/growth-in-minority-executives

Elsie Boskamp. "63 Diversity in the Workplace Statistics [2022]: Facts You Need To Know," Zippia.com, December 20, 2021, www.zippia.com/advice/diversity-in-the-workplace-statistics

Ann Brown. "I'm Not An Assistant, I'm The Owner: How Women Of Color Experience Unconscious Demotion," Madamenoire.com, March 13, 2018, madamenoire.com/1017378/no-im-not-an-assistant-im-the-owner-how-women-of-color-experience-unconscious-demotion-at-work/- :~:text=According to many studies, this is a common,others and their career status based on race

Leverage Your Privilege

"When we identify where our privilege intersects with somebody else's oppression, we'll find our opportunities to make real change."

—Ijeoma Oluo

***

Thanks to an executive order by John F. Kennedy in 1961, Affirmative Action began to shift the tide for people of color in the workplace. Affirmative Action prohibited discrimination in the workplace based on race, disability, gender, ethnic origin, and age. Sixty-plus years later organizations continue to struggle to create and sustain diverse workforces. Discrimination remains alive and well as cultural bias toward White supremacy persists. Consider the research cited in the previous activity stating that the workforce in the United States is comprised of 78 percent of White people. Something is clearly wrong when according to the 2020 census data, White people represent 57 percent of the total U.S. population. Even when companies manage to improve representation, people of difference are made to feel like outsiders through assimilation, hostile environments, and prejudicial policies and systems, thus creating unearned, unacknowledged rights, benefits, and opportunities for the dominant culture while systemically oppressing everyone else.

Diversity practices focus on attracting and hiring members of under-represented groups, while equity and inclusion is the practice of retaining them. The intent behind diversity, equity, and inclusion efforts is for employers to ensure that everyone can fully participate through fair practices and equal opportunities to advance. The privilege of the dominant culture is something that is highly visible and obvious to people of difference as we face life-diminishing, life-threatening, or life-ending circumstances regularly that our White counterparts only hear about. There needs to be a better understanding of what White privilege is and how it impacts someone's life, both inside and outside the workplace. Those with privilege must decide whether they want to do the work to dismantle systems of oppression and make the conscious choice to no longer perpetuate them. I find that White people are aware of the disadvantages of being Black, Hispanic, Asian, or indigenous but often fail to see the advantages of being White until it's called to their attention like with the highly publicized murder of George Floyd in 2020. On a more personal level, my White friends who have children with Black and brown people have been truly awakened to their privilege, as they realize the perils that their brown-skinned children will face based on the color of their skin. Some have even been ostracized for their decision to marry outside their race. I'm appreciative and yet amazed by those who privately show support and empathy as doing so publicly could cost them dearly. However, many White people will argue that White privilege is not real and has nothing to do with their success. After all, they worked hard for everything they have. That said, non-White people work harder for the same level of achievement and must navigate systems designed to disadvantage us in hopes that we can overcome them. I was once told by an employer that I could not be considered for promotion to a sales manager position because they had reached their quota. So, I asked whether there was a quota for White people, and the reply was no. I have a Latina colleague in a middle manager role who was talked down to by a White, male intern. It was clear to her that he felt entitled to do so because of her gender and race. She never heard him speak in that condescending tone when speaking with White people, and he was reverential when in the presence of leadership. Sometimes White privilege can be more subtle like when White people are given the benefit of the doubt and non-Whites are judged more harshly when mistakes are made.

While systems have been built to create advantage and privilege for White people, everybody else has some level of privilege based on class, ability, education, or sexual orientation that can be utilized to help others. Consider seeking opportunities to support gay rights, advocate for more accessibility for individuals with disabilities, or mentor individuals to help elevate their career. Use privilege to inform conversations, become more empathetic, and connect with people on a more personal level. Privilege empowers us to speak up and take action on behalf of those without it. Champions of inclusion are aware of our privilege and leverage it to provide life-enriching, life-changing, or life-saving experiences to others.

Actions

Know Your Privilege

Reflect on the parts of your identity that you think about least. Do they afford you safety and make you feel valued and welcomed in every situation throughout your day to day? Consider going to a conference. Are you mindful of pronouns people use when they refer to you, will skin color impact an opportunity or experience, or must you consider an accessible route from point A to point B? Answers to typical questions like these can help you identify your privilege. Notice how it impacts your day. This exploration can enhance your empathy for those who don't share the same privilege.

Leverage Privilege Regularly

Privilege exists in many situations, within different lifestyles, and among every demographic. It's important that we understand that not everyone has equal privilege. Educate yourself on the challenges of individuals and communities that lack privileges that you may have based on your identity, lifestyle, education, etc. Seek opportunities to show up and make an impact—from confronting bias to raising awareness of issues and introducing a point of view that may not organically emerge in discussions at work. Be willing to share time and resources. Ask yourself, "What am I willing to do to use my power or privilege in a way that improves someone's lived experience today?"

Raise Awareness Through the "Privilege Walk"

You may be aware of or even participated in a "privilege walk"—an activity to help us recognize how power and privilege affects lives even when we are not aware it is happening. It's centered around the ability to identify both obstacles and benefits that everyone experiences. After watching the video *What is privilege?*, reflect on the questions asked to identify your privileges or lack thereof. Consider the role they played in your current state of life. Connect to any emotions you have during the reflection to build empathy or understanding. If you're feeling inclined, organize a privilege walk. The Transfer Leader Institute offers a great resource: www.eiu.edu/eiu1111/Privilege%20 Walk%20Exercise-%20Transfer%20Leadership%20Institute-%20Week%204 .pdf#:~:text=Instructions%3A%20Everyone%20will%20stand%20in%20 a%20horizontal%20line,step%2C%20they%20have%20the%20option%20 to%20remain%20still.

Action Accelerators

- **Netflix documentary**: *Hello, Privilege. It's Me, Chelsea.* Here's the trailer: www.youtube.com/watch?v=tPsLcrVlwt4

- **Checklist**: "White Privilege: Unpacking the Invisible Knapsack": psychology.umbc.edu/files/2016/10/White-Privilege_McIntosh-1989.pdf

- **YouTube.com**: *What is privilege?*: www.youtube.com/watch? v=m3AMhtHPKNk

- **JenniferBrownSpeaks.com**: *Being a Better Ally: How to Use Your Privilege To Create More Inclusive and Engaging Workplaces*: www .stitcher.com/show/the-will-to-change-uncovering-true-stories-of-diversity-and/episode/e53-being-a-better-ally-how-to-use-your-privilege-to-create-more-inclusive-and-engaging-workplaces-59433850

- **Huffpost.com**: "6 Things White People Say That Highlight Their Privilege," by Kelsey Borresen: www.huffpost.com/entry/things-white-people-say-highlight-privilege_l_5edeafafc5b637b8 7e22cee0

- **YouTube.com**: *Census shows US more diverse than ever as White population declines,* CNN: www.youtube.com/watch?v=Gz_fc93YYkg&t=26s

- **YouTube.com**: *White people are Not Ignorant about Racism/WS,* Jane Elliott: www.youtube.com/watch?v=UBHJpWoFDwo

Support the Firsts, Fews, and Onlys

"I sit on a man's back choking him and making him carry me, and yet assure myself and others that I am sorry for him and wish to lighten his load by all means possible . . . except by getting off his back."

—Leo Tolstoy

***

As organizations continue to diversify the workforce, there is bound to be the "firsts, fews, and onlys" (FFOs). They are easy to recognize because there is no one else in the organization who closely resembles them. The new customer service representative is the one individual on the team who has an artificial limb (first and only), your department is comprised of 50 people and only 6 percent, or three of them, are Black (few), or the VP of sales is lesbian and the only LGBTQ individual in leadership ranks. This is not an unusual experience for those navigating workplaces and industries known for being predominantly cisgendered, heterosexual, absent a physical disability, White, and male. FFOs are largely members of historically excluded groups and often experience feelings of isolation and hyper-visibility in the workplace, which can become intensified at the intersection of identities. For instance, say the VP of sales is not only gay but is also African American and female —representing three marginalized identities. The hyper-visibility creates greater pressure to perform without mistakes and results in being held to higher standards of productivity while conforming and assimilating to be seen as competent. As a member in the FFO club myself, I can speak from experience that there is a high likelihood of having abilities challenged, work overscrutinized, enduring microaggressions, and being sexually harassed. Everyone who falls outside the FFO club tends to be oblivious to the inequities and unfair treatment that they perpetuate. There is no one around who even recognizes, and therefore can speak up against, biased, racist, and sexist behaviors, and those who are aware are usually too afraid to do anything about it. For them, going against the grain is too risky. When there is no one else who can help educate the members of the dominant culture who "just want to learn" or "just want to help," FFOs are left with the emotional labor of constantly explaining and answering questions about what it's like to be them. They bear the weight of being the sole representative of their race or culture for which the entire group will be judged, or they are appointed the face of diversity for the organization. FFOs will soon find themselves in a perpetual state of educating co-workers or performing on behalf of their race, rather than just being who they are and doing the job that they were hired to do to the best of their ability. The heavy load of "educator" and "performer" zaps energy and takes a toll over time. Stress, anxiety, burnout, guilt, or shame may result from the painstaking work of these unwritten and unspoken roles. FFOs are constantly negotiating with themselves on teaching people about their experience as a person of difference while staying positive and being a valuable team player.

Sadly, many FFOs, especially those in senior ranks, are set up to fail as the plan was for them to never succeed; rather, were a quick-and-easy performative measure to evidence diversity—a tactic employed by organizational leaders to give the illusion of inclusion. In earnest, it is an attempt to avoid public backlash for not actually being diverse, equitable, and inclusive. Absent a sincere commitment to diversity, equity, and inclusion, FFOs are tokenized. *Tokenism*, a symbolic effort to be inclusive of marginalized groups to give the appearance of diversity efforts, is nothing new. It became popular in the late 1950s in the midst of racial segregation. It is merely a means of acknowledging the lack of diversity without putting in the work to actually create and sustain it.

Our ability to support FFOs in the workplace begins with the *why* established in Activity 2, "Connect with Your *Why*, Find Your *Why Now*." Our *why* will inform our actions, interactions, decisions, and ideally build the courage to act, the ability to connect, and the awareness to empathize. We must challenge ourselves to interrupt biased thinking and disrupt stereotypical beliefs. It can be challenging to know where to begin an authentic journey toward inclusion, especially when we, ourselves, are the first, one of the few, or the only ones doing so. Courage helps us to overcome the reluctance of putting ourselves out there for fear of getting it wrong. Connecting will establish the trust needed for authentic dialogue as well as discovering common interests and similarities. Empathy creates the ability to appreciate the emotions and experience of others, thus building a bridge across the divide. Inclusive actions are intentional, whereas exclusive ones go unnoticed as part of the status quo. Here are a few ways to support the FFOs in your organization:

- Recognize them as an individual, not the group they represent.
- Acknowledge the value they add to the team with their skills and expertise.
- Embrace their unique perspectives and contributions as opportunities worth exploring rather than problems to be dismissed.
- Help get them acclimated to the team and the team acclimated to them by focusing on commonalities, not differences.
- Educate yourself on the issues and challenges of marginalized groups, rather than asking FFOs to do the work for you.
- Share common practices like team norms and unwritten rules.

Champions of inclusion are sensitive to the challenges of firsts, fews, and onlys; are the first to welcome them to the team; and work toward creating an inclusive work experience where thriving is the norm.

Actions

Be the Welcome Wagon

Help firsts, fews, and onlys feel like part of the team from day one or even if they have been there for a while and you have yet to get to know them. It's never too late to introduce yourself and make someone feel welcomed. Practice inclusive habits and work toward connecting (refer to Activities 3 and 4). Help the team make space for them and become an ally (refer to Activity 13, "Aspire to Be an Ally.")

Course Correct Where You Can

Inclusion journeys are a marathon, not a sprint. It's going to take time and practice to learn the skills to be more inclusive and connect across differences. FFOs primarily exist in organizations lacking a real diversity, equity, and inclusion strategy. While most of us in the workforce do not have the positional authority to drive change, we can change our perspectives and the way we interact with people who are unlike us. Own the part you play that causes people of difference to feel unwelcomed and make a personal commitment to do better. Consider the FFOs in your organization, and ask yourself, "What can I do today to enhance their experience?" Do it again tomorrow.

Realize the Value of Being Inclusive

Let's make a movie and you're the star. Imagine arriving at a family celebration that for weeks you've been looking forward to attending. Everyone is expected to be there. There will be folks you've known all your life and others you have yet to meet. There's the promise of entertainment and all of your favorite foods. (Pause. Write down your feelings and expectations.) As you enter the room, you can see everyone laughing, smiling, and enjoying one another's company. You stand there for a few minutes taking it all in, and no one acknowledges your presence. Seemingly it's no big deal; everyone is having a fun time and too busy to notice you. You spot some folks playing a card game and sit down to be dealt into the next hand. Instead, they bypass you, and the game continues around you. You didn't get a single card. (Pause. Write down your feelings and expectations in playing the game.) As you stand alone, your aunt Mary is coming your way, and you expect to take a few minutes to catch up. Instead, she points out that you are blocking the beverage cooler, asks that you please move, collects two beers, and then walks aways without another word. A few moments later, you notice a group over by the pool, grab a beverage from the cooler and walk over to join them. They notice you approaching and quickly disperse. (Pause. Write down your feelings and expectations of the encounter.) By now you are wondering why you're getting the cold shoulder from everyone. Has someone spread lies about you? Did you do something that has pissed off the entire family and didn't realize it? Perhaps this is some crazy joke that no one has let you in on. In an attempt to get answers, you notice your cousin Joe across the way. He is in your line of sight. You watch him as you call his number; he declines your call, puts the phone back in his pocket, and resumes his activity. Your movie has just ended. (Pause. Write down your feelings and expectations from Joe.) Look at your list of emotions experienced in just trying to fit in and connect. You have just gotten a glimpse into a workday in the life of an FFO. Reflect on how your experience would be impacted if one person finally approached, greeted you by name, and said that they were happy that you could join. (Pause. Write down your feelings.) Now you are positioned to see the value of inclusion. Be the person who shows up.

Action Accelerators

- **YouTube.com**: *The Power of Onlyness*, Nilofer Merchant: www.youtube.com/watch?v=wUOd-RRfLBA&t=1085s

- **Podcast**: JenniferBrownSpeaks.com, E122: Building Successful DEI Initiatives, The Importance of Psychological Safety: jenniferbrownspeaks.com/?s=Building+Successful+DEI+Initiatives+|+The+Importance+of+Psychological+Safety

- **Book**: *Confessions from Your Token Black Colleague* by Talisa Lavarry: yourtokenblackcolleague.com

- **Forbes.com**: "The Dangers of Mistaking Diversity for Inclusion in the Workplace," by Dana Brownlee: www.forbes.com/sites/danabrownlee/2019/09/15/the-dangers-of-mistaking-diversity-for-inclusion-in-the-workplace/?sh=734fdbc74d86

34

Cultivate Acceptance

"Our uniqueness, our individuality, and our life experience molds us into fascinating beings. I hope we can embrace that. I pray we may all challenge ourselves to delve into the deepest resources of our hearts to cultivate an atmosphere of understanding, acceptance, tolerance, and compassion. We are all in this life together."

—Linda Thompson

***

Now, more than any other time in history, people are feeling more comfortable in expressing themselves for who they truly are. The reality or the pursuit of living life authentically can be life changing for anyone, but especially those in the transgender community. That said, it's quite likely that we personally know someone who identifies as transgender. Maybe we're aware of who they are or have yet to realize it. Perhaps they are the barista at the coffee shop, your dentist, or your manager. According to the Williams Institute, 1.4 million adults identify as transgender in the United States. Notable members of the transgender community include Laverne Cox, Emmy-nominated actress; Layshia Clarendon, WNBA player; and Martine Rothblatt, CEO of United Therapeutics.

Many successful transgender professionals will tell you that a primary reason for their success lies in their ability to live their lives genuinely. The alternative of staying closeted is too overwhelming and mentally exhausting. Yet, the transgender community is probably the most ostracized, most stigmatized, and most discriminated against community in the world. Often, they are rejected by family, bullied, and harassed by co-workers, as well as shamed by mainstream society and the education system, making life a living hell for many. Transgender people are individuals whose gender identity does not align with the one assigned at birth (see Figure 34.1). They may be someone who undergoes hormone replacement therapy and/or surgical interventions to achieve their true selves. Others may transition without the assistance of healthcare professionals and instead opt for changes in everyday dress, mannerisms, or speech. Everyone's transition journey is different.

Transgender individuals experience workplace discrimination and harassment at higher rates than gays and lesbians. Although there is an increasing awareness of the challenges faced by transgender individuals in the workplace, many organizations are not adequately prepared to create the policies and build the culture needed in support of transgender employees. In such a climate, the lived experience of the "first" or "only" can be traumatic when staff has not been trained or educated on inclusive practices, the LGBTQ community as a whole, and the nuances of the transgender community. For trans individuals, making the difficult decision to transition from presenting publicly as a man or woman to the gender they truly identify with requires a supportive environment to help avoid devastating instances of abusive or inappropriate treatment at the hands of leaders and co-workers. Encountering unchecked bias, being confronted by stereotypical beliefs, and constantly explaining oneself fosters a toxic environment. A post in Newsweek.com states,

> "Of the 90 percent of transgender workers who faced discrimination at work, about a fourth were forced to use restrooms that did not match their gender identity, were told to dress, act and present as a different gender from their own in order to keep their job, or had a boss or coworker share private information about their transgender status without their permission."

This type of discrimination and hostility forces them to hide their identity, can lead to increased absenteeism; lack of commitment and motivation to do their best work; or, worse, contemplations of suicide.

LGBTQIA Defined

Based on the LGBTQ+ Vocabulary Glossary of Terms at TheSafeProject.com
https://thesafezoneproject.com/resources/vocabulary

L Lesbian	**G** Gay	**B** Bisexual
noun & adj. : women who are primarily attracted romantically, erotically, and/or emotionally to other women.	1 adj. : experiencing attraction solely (or primarily) to some members of the same gender. Can be used to refer to men who are attracted to other men and women who are attracted to women. 2 adj. : an umbrella term used to refer to the queer community as a whole, or as an individual identity label for anyone who is not straight.	1 noun & adj. : a person who experiences attraction to some men and women. 2 adj. : a person who experiences attraction to some people of their gender and another gender.
T Transgender	**Q** Queer	**I** Intersex
1 adj. : a gender description for someone who has transitioned (or is transitioning) from living as one gender to another. 2 adj. : an umbrella term for anyone whose sex assigned at birth and gender identity do not correspond in the expected way (e.g., someone who was assigned male at birth, but does not identify as a man).	1 adj. : an umbrella term to describe individuals who don't identify as straight and/or cisgender. 2 noun : a slur used to refer to someone who isn't straight and/or cisgender. Due to its historical use as a derogatory term, and how it is still used as a slur in many communities, it is not embraced or used by all LGBTQ people.	adj. : term for a combination of chromosomes, gonads, hormones, internal sex organs, and genitals that differs from the two expected patterns of male or female. Formerly known as hermaphrodite (or hermaphroditic), but these terms are now outdated and derogatory.
	**A** Asexual adj. : experiencing little or no sexual attraction to others and/or a lack of interest in sexual relationships/behavior. Asexuality exists on a continuum from people who experience no sexual attraction or have any desire for sex, to those who experience low levels, or sexual attraction only under specific conditions.	 PRIDE RELATIONSHIP PEACE RIGHTS ACCEPTANCE FREEDOM LOVE EQUALITY ALLY

Figure 34.1   LGBTQIA glossary

Our role in the workplace is to become more aware of the inequities, stop perpetuating them, and support our colleagues through them as best we can. Supporting is far easier said than done. Unless you are a trans individual or have loved one who is, it's impossible to understand, let alone empathize, without first educating ourselves. Many well-intentioned, individuals who are non-LGBTQ become self-conscious, and in attempts to support find themselves over catering and tiptoeing around in an effort not to offend, which further marginalizes trans co-workers by being singled out. When the extra care or concern isn't extended to everyone, people notice. Champions of inclusion acknowledge that trans individuals just want to be treated and deserve to be treated with the same dignity and respect as everyone else. Fostering a culture of acceptance is simply the right thing to do. Someone can leave work today as a man named Jack and return tomorrow as a woman named Jill when acceptance is unconditional. We don't have to like it. We don't have to pretend to like it. We can *always* be respectful. Our intentional efforts to connect and build relationships with our trans teammates, working to understand the issues and challenges of transitioning as well as those of being transgender, is the best way to show support. When Heidi, a female to male transgender person whose pronouns are *they/them* worked as a nurse practitioner, they were very open about their transition. Many colleagues distanced themselves as Heidi became more masculine, while others embraced it. Supporters were mindful to use their name—Hunter—and *they/them* pronouns and apologized when they made mistakes. They showed up as allies when they did not entertain gossip as well as educated folks who were open and wanted to learn more. For Hunter, their support was priceless. Invest the time to understand the trans lived experience, work to create a safe space to learn, discuss, and explore one another's misconceptions and anxieties. This is how we build cultures of acceptance, belonging, and inclusion.

Actions

Go Beyond Allyship—Be a Friend

True and authentic friends are one of life's greatest gifts. Many of us who are heterosexual and cisgendered don't have friends in the transgender community, and we're missing out. As we connect through allyship, aspire to become a friend.

Respect Identity

Knowing what to do is just as important as knowing what not to do. Get to know trans colleagues and treat them as they want to be treated (refer to Activity 38, "Reconsider the Golden Rule,") not as you would want to be treated. Use their pronouns and the name they have chosen in written and verbal communications. We all slip up sometimes. Just apologize and don't make a big deal about it.

Learn the Facts

The transgender community is one of mystery to nontransgender individuals. The many unanswered questions lead to wrong assumptions, false conclusions, and inappropriate behavior. Dispel stereotypes and misconceptions by learning the facts. If your organization has an LGBTQ or trans affinity group, join as an ally. In addition, review this library of resources to get started:

- www.glaad.org/transgender/resources
- transequality.org/about-transgender
- pflag.org/transgender

Observe Transgender Remembrance Day

Transgender Day of Remembrance (#TDOR) occurs annually on November 20 and is a day that remembers those transgender people who lost their lives in acts of anti-transgender violence. There may be gatherings local to your area like dedicated church services, marches, art shows, food drives, or film screenings. Consider contributing to a not for profit in support of trans individuals like the Jim Collins Foundation, the Transgender Law Center, or the National Center for Transgender Equality.

Action Accelerators

- **Everfi.com:** "Transgender Issues in the Workplace: Bathroom Access, Workplace Abuse, Hiring Discrimination," by Scott Raynor: https://everfi.com/blog/workplace-training/transgender-issues-workplace-abuse-hiring-discrimination

- **Insider.com:** "9 Problematic Phrases You May Not Have Realized Are Transphobic," by Canela López Nov 25, 2020, `www.insider.com/phrases-you-should-never-say-to-transgender-people-2020-11`

- **McKinsey.com:** "Being Transgender at Work": `www.mckinsey.com/featured-insights/diversity-and-inclusion/being-transgender-at-work`

- **Book:** *Before I Had the Words* by Skylar Kergil: `www.beforeihadthewords.com`

Sources Cited

WilliamsInstitute.Law.UCLA.edu: "Subpopulations – Transgender People," `williamsinstitute.law.ucla.edu/subpopulations/transgender-people`

Christianna Silva. "Almost Every Transgender Employee Experiences Harassment or Mistreatment on the Job, Study Shows," Newsweek.com, November 29, 2017, `www.newsweek.com/transgender-employees-experience-harassment-job-726494`

INTERVIEW: TELLING IT LIKE IT IS. . .TO GET WHERE WE NEED TO GO

Move Beyond Allyship, Be a Friend

Meet Ayden. He is a vibrant member of the trans community and unapologetically comfortable with who he is now. He, like most of us, believes himself to be likable, caring, and fun loving and would never intentionally harm or cause anyone to feel unsafe. Like everyone else, he wants to be treated with dignity and respect. While managing a café, he began his transition from presenting as Amber to presenting as Ayden, and up until that point, the work environment was good. As he changed, the environment changed with him, and not for the better. It became apparent that that was not the place to introduce his true self. The harassment and never-ending transphobic comments spoke volumes and sent a clear message that he was no longer welcomed there. So, he took his excellent people skills to a new organization and launched a career in sales at a global e-commerce company.

Getting comfortable can be a challenge in new environments, and this one was no different. In fact, people became curious right away. During training, a fellow trainee flat out asked in front of the entire class, "Hey, are you transgender? You always sit down when you go to the bathroom!" The manner and tone in which it was asked was as casual as saying "Hey, what's your name?" It was evident to Ayden that his co-worker felt that it was perfectly okay to ask something so private in a public forum. He had no clue of the impact of what he had just done. But let's be clear, even asking that question in private is never okay. It's like asking, "Hey, you got diarrhea? I noticed you stank up the bathroom really bad" or "Hey, are you on your period? You are really moody today." Making statements or asking intrusive personal questions creates a hostile work environment, and the humiliation has lingering effects. Aghast, Ayden kept his composure, absorbed the insulting blow, and responded "yeah" in a no-big-deal kind of way. Mentally, he was thinking, "Whatever, I'm the new kid. Let's not make this a big deal and let's keep it moving." Although this happened several years ago, the details remain fresh in his mind as if it occurred yesterday. Ayden suspects that the person who outed him that way never gave it a second thought, and the actions of that day will be forever etched in Ayden's mind as a traumatic experience. Even though Ayden learned later that several individuals went to human resources (HR) on his behalf, the trauma from his previous employer and being a new hire drove his decision to stay quiet and not address it with HR. His desire was to start fresh, be judged on his performance, and simply be known as Ayden and not "Ayden the trans guy."

Eventually Ayden left for greener pastures and continued his sales career at an international luxury retailer. In typical Ayden fashion, he was open about being trans from the online application process and right into the interview with HR and the hiring manager. His transition continued, and he stepped onto the sales floor as AJ. He chose to wait to come out as Ayden to co-workers when he was six weeks into hormone replacement therapy. By now, the effects were becoming noticeable. The need to manage their reactions and prepare them for the physical changes that they would witness each day became paramount for his psychological safety. His closest allies were two straight male co-workers who treated him like one of the guys and always introduced him as "This is my boy, Ayden." Finally, for the first time, he belonged, and it was lifesaving. Their friendship was a haven at work. For Ayden, allyship was just part of being a friend. His new connections demonstrated friendship with their authenticity, as well as acts of care and support. They treated him the same as they did everyone else. They spoke up and helped make work safe for their friend, and he is forever grateful. Their friendship has continued to flourish over the years. Nothing feels better than being accepted for who you are.

Be the Calm in the Storm

"Compassion is not a virtue—it is a commitment. It's not something we have or don't have—it's something we choose to practice."

—**Brené Brown**

***

Organizations place value on innovation, productivity, and efficiency. It creates a competitive advantage and drives profits. Before individuals start the first day on the job, most must answer this question during the interview process: "Why do you want to work here?" If we've done our homework, we know what the company does, what they plan to do, and how our experience, skills, and competencies can help them meet goals and carry out their mission. This is our value proposition. Once hired, the expectation is to perform at levels high enough to carry out responsibilities and meet or exceed goals. Anything less can bring forth negative consequences—such as poor performance reviews, 90-day performance improvement plans, probation, suspension, or job loss. No one is perfect, and everybody is allowed a bad day or two here and there. It's when we start to have repeated strings of bad days that our value to the organization starts to diminish. I began my career in an era where one was expected to be fully at work as scheduled and get the job done no matter what. If your child was too ill to attend school, the expectation was to make other arrangements and get to work. Impacted by the Rodney King verdict, the Matthew Shepard killing, race riots, and civil unrest? Please don't talk about it in the workplace; it's inappropriate and diminishes productivity. Accomplishing the mission of the organization was priority #1 and ingrained in employees at the expense of the human condition and real-life circumstances. It didn't matter when bosses demonstrated little regard for staff members—usually women—with a sudden child-care problem. Never mind that one may be grieving or traumatized by systemic and institutionalized acts of racism and violence. Coming to work meant *being* at work. Grieving, discussing, and the processing of it all was done at home, at church, or with community, and absolutely, positively not in the workplace.

The majority culture has always defined what work is and what it looks like within an organization, and it starts in the C-suite. When there is a people-first mindset, the policies, practices, and benefits are designed to help employees thrive, even in the worst of circumstances. Operationalizing them has its challenges, but at least there are resources that can be leveraged. By contrast, the job-first culture prioritizes get-ting the work done with little regard for the human condition and how circumstances outside the workplace impact what goes on inside the workplace and informs how we interact with one another. Thus, putting it on the individual to "just get over it" makes folks feel even more lost, broken, or disconnected. The tactic of gaslighting, manipulating someone to the extent that they question the validity of their own thoughts, feel-ings, or perception of reality, is probably the most damaging and most effective at creating feelings of exclusion. We now find ourselves in an era where more than ever our voices can be heard—even when folks are trying hard *not* to listen. We are empowering ourselves as well as one another to incite real change in the workplace where everyone has an equal chance to thrive each day. We are reacting in real time to incidents impacting us as individuals and society as a whole.

Occurrences of racism, bigotry, hatred, and prejudice represent pain-ful and traumatic experiences for women, Black, Brown, and queer co-workers both inside and outside the workplace. People are being killed literally and figuratively outside the workplace through acts of violence and laws and legislation allowing elected officials to make decisions on our behalf that we would not otherwise make on our own. As a result, people are suffering inside the workplace. How we respond on an individual level can either exacerbate the situation or help remedy some of the anxiety and stress. When we respect, listen, value, and speak up for each other, nurturing cultures emerge, and we demonstrate value for humanity in all its complexities. Whether someone is navigating organizational trauma or vicarious trauma, being consistent in our actions will influence culture for good and have a lasting impact. As affected co-workers work to heal and/or process emotions of anger, fear, or helplessness, our empathy and acts of kindness become essential. Often, we have no idea of the lingering effects of our words and deeds when folks are hurting. I was in the workplace at the time of the Rodney King beating and subsequent 1992 acquittal of four Los Angeles police officers for one of the most savage beatings I had ever seen on television news. I was horrified at the sight of the images and remember feeling sick. I was also in the workplace when the murder of George Floyd received worldwide attention, just as the Rodney King incident. The difference was like night and day. Outside the workplace, Rodney King was on everyone's mind. We lamented racist acts and the unjust justice system. The community came together and supported one another. Once in the workplace, there was radio silence. It was as if it never happened. When I asked White co-workers their thoughts (I was the only person of color on my floor—so there were only White people to engage in conversation), I would either get a nonchalant response of "Yeah, that was too bad" or "That man got what he deserved." No one cared to talk about it, asked me how I felt about it, or even showed concern for Rodney. I was baffled by the dispassionate response from co-workers. Could it be that they actually did not care? Would their reaction have been more passionate had Rodney been one of their own? It took me weeks to get back to my levels of productivity prior to the beating. When the verdict came in about a year later, I did not mention it at work, but the outrage I felt followed me there. Leaving it at the door was impossible, so I masked my feelings and behaved as if nothing happened—as I was expected to do. My co-workers made it clear that my feelings about what happened to Rodney and its personal impact were not their concern. This was not the response of just my employer, but my friends and family members experienced the same attitudes at their places of work. By contrast, when the videos of the George Floyd murder hit the airwaves, of course, it was top of mind with family and friends. Yes, we were still lamenting, 28 years later, acts of racism and the unjust justice system that may just acquit the police officer accused of his death. But the difference this time was that in many workplaces—not all—the same conversation was happening. Co-workers were checking in on each other, we were talking about it in meetings without consequence or criticism, we had venues to hold constructive conversations about race, and we were educated on and encouraged to use employee mental health benefits and the like. We felt safe enough to share personal stories and perspectives. The grace meant a lot. It reduced stress levels and helped me maintain productivity with less struggle. Not having to check my emotions at the door fostered stronger connections with co-workers and gave me an outlet as I struggled to manage getting the work done while processing the pain and sadness. Sadly, hate crimes continue, and we continue to arrive in the workplace traumatized by them and other things beyond our control. It's important to realize that we are not all affected equally by systems of inequity. Champions of inclusion value humanity and show up to be the calm in the storm. We let folks know that we are available to listen and help where needed. We accept our role in shaping the environment. Anything less just does more harm to those who are hurting. Knowing that we're not alone in the workplace and that there are safe spaces makes us feel included and allows time to heal, to be the best version of ourselves, and to deliver on our value propositions to organizations.

Actions

Acknowledge What's Happening

The first step to showing up is acknowledging that something has happened, even if we have not been personally impacted by it. We can't pretend that everything is fine when it's not just because it did not affect us. We can't ignore the impact of horrific events outside the workplace or acts of oppression in the workplace. We don't need to be in a leadership role or position of authority to demonstrate compassion and empathy. Saying something or doing something lets those hurting know that we are there. Send an email, send a text, make a phone call, and make it authentic. Lift some of the burden of work by offering to partner on a project or meeting, cover a shift, extend a deadline if possible, or do something that provides a little space and grace for one to process and heal. This isn't the time to look on the bright side. Sometimes reality sucks.

Remember to Check in Periodically

Folks don't heal from traumatizing events in a day, a few weeks, or even months sometimes. Everyone processes hurt differently. Check in with people from time to time. Don't make it awkward, and don't resurface the traumatizing event by including it in the interaction. Just ask a heartfelt "How are you doing?" and listen to what is said and unsaid. Let folks know you're thinking about them. It could be a text, email, or instant message: "Hey, you were on my mind today. Do you need anything?" Be prepared for people to respond with honesty. When people trust us enough to be honest, we must respect them enough to listen. When promises are made to follow up or do a particular thing, keep the promise.

Learn the *Why* Behind Triggers

When traumatic events happen outside the workplace or acts of oppression happen inside the workplace, individuals and groups are affected at different levels in different ways. Don't expect folks to explain the significance of traumatizing events. For example, if you don't understand why the insurrection on January 6, 2021, outraged Black people on a deeper level than White people, don't expect to have it explained to you. Listen to the perspective of Black news commentators or social media influencers. If you don't understand why nonbinary colleagues feel singled out and discriminated against in the absence of unisex restrooms, don't ask—do your research. Increase your understanding of movements to include Times Up, Black Lives Matter, Me Too, and Stop Asian Hate. The enhanced perspective and understanding will allow us to be more supportive.

Action Accelerators

- **History.com**: "This Day in History, March 3, 1991: LAPD officers beat Rodney King on camera": www.history.com/this-day-in-history/police-brutality-caught-on-video

- **Cigna.com**: "Supporting Coworkers and Others After a Traumatic Event": www.cigna.com/individuals-families/health-wellness/how-to-support-others-after-trauma

- **YouTube.com**: *Brené Brown's Life Advice on Emotions Will Leave You Speechless*, www.youtube.com/watch?v=-aU6T-ed1eI

- **Ted.com**: *What we can do about the culture of hate*, Sally Kohn, TEDWomen 2017: www.ted.com/talks/sally_kohn_what_we_can_do_about_the_culture_of_hate

- **Business.kaiserpermanente.org**: "The new workplace is trauma-informed," by Cosette Taillac, LCSW: business.kaiserpermanente.org/insights/covid-19/trauma-informed-workplace

Acquire an Education on Race and Racism

"The next time you come across something that sounds too outrageous to be true, know that your instinct may be correct. Ultimately, it's up to all of us to think critically, do our research, and determine for ourselves whether a source can be trusted."

—**Rachel Hartman**

***

Musical genius Stevie Wonder was onto something when he sang this lyric from the song *Superstition*: "When you believe in things that you don't understand, then you suffer. Superstition ain't the way." We've all been trapped by beliefs of others that we didn't fully understand. Whether superstition, conspiracy theories, critical theories, advice from experts and self-proclaimed experts, even guidance from parents and teachers can leave us confused and uninformed. I wonder how many of us are unhappy in careers right now based on the counseling of people we respect because we didn't ask enough questions and approach the information with a curious mindset. When Islamic suicide bombers took their own lives and the lives of others on September 11, 2001, they may have been convinced that their martyrdom would result in sexual delights in the afterlife. Our perspectives are one-dimensional when we don't seek wisdom from various sources and more importantly sources that fall outside our circle. Otherwise, we simply believe and act accordingly in an echo chamber. Most White people that I encounter tend to underestimate the depths of racism and believe it to be a thing of the past in the United States citing that we elected Obama twice. They are oblivious to racist systems and refuse to seriously examine the perspectives of people of color as they do their White counterparts.

Race and racism in America go beyond the acts of individuals who consider racism to be about hating, disliking, or harming others on the basis of race. It's a much deeper problem. The barriers to equality for Black and Brown people in America are rooted in systemic and institutionalized racism that has been alive and well for centuries and that shapes decisions that relegate people of color to the lowest tiers of society. The system is designed to disenfranchise anyone not in the dominant culture. Our journey on the road toward inclusion requires a transformative process where we shift our thinking, attitudes, and behaviors to the extent that we can achieve equality for individuals of historically marginalized groups. We intentionally strive to create equal access to opportunity, wealth, and privilege. It's about the elimination of oppression and addressing racism to create full and equal participation of all groups in our society and places of work where the distribution of resources is equitable. It's about building a fairer and more socially inclusive world that is physically and emotionally safe and secure. It's about acknowledging that one race is no more superior than another. Just imagine the impact on the world and the workplace when everyone has equal opportunity to fully partake. Have you ever stopped to wonder why these topics are still relevant despite decades of work to drive change? Why must we be more intentional in the first place? What got us here? I can assure you that it's not by accident. As stated in a previous chapter, no one is born racist. Racism is learned. Thus, it can be unlearned.

My efforts here are to awaken folks to the fact that just because this is where we are, it does not mean we have to stay here. We must ask ourselves, "Why do I want to remain in long-held values and beliefs that cause harm to so many? How would I feel if it was happening to me, and how would I effect change if this was my lived experience?" The answers to these questions provide a moment for introspection, a chance to seek a different point of view, and have our consciousness confronted. We may find that a new way of thinking about race and racism lies in critical race theory (CRT). It's the wild, wild west out there when it comes to opinions on the topic. The use and misuse of the term combined with guilt and shaming tactics set us up for inevitable conflict based on misrepresentation of the facts, flawed interpretations, or both. If we are truly serious about being part of the solution, it's important to be mindful of who we trust for factual information. Mindboggling is an understatement in my effort to understand how folks could be so vehemently opposed to something that they know nothing about, or haven't attempted to get the facts for themselves, yet they are yelling and screaming to high heaven about how wrong something is and how it impacts them and their families. To thrive in this life, we've got to learn to discern fact from fiction and opinion. Acting on false beliefs and false ideologies of those who want to control the narrative causes harm and perpetuates ignorance. I may never understand why any sensible person would accept the opinion of another as fact and act on it without first investigating. But nonetheless, it happens! Consider when President Donald J. Trump advised Americans to drink bleach to protect themselves against the covid-19 virus. The Center for Disease Control (CDC) reported that "the daily number of calls to poison centers increased sharply. . .and that bleaches accounted for the largest percentage of the increase." While this is truly sad, the individuals who followed the advice could have avoided harmful or life-threatening effects had they paused to consider that the source of the information was not from a medical professional. Even if they believed the source to be trustworthy, as I am guessing they did, why not investigate? Why not ask your doctor or even read the label on the bleach bottle to validate the message and make an informed decision before acting? The label on the back of my Clorox Bleach bottle literally says "DANGER. CORROSIVE. Causes irreversible eye damage and skin burns. Harmful if swallowed." Further it says "IF SWALLOWED, have the person sip a glass of water if *able* to swallow. Do not induce vomiting unless told to do so by a poison control center or doctor." I am speaking only for myself when I say that the label warning that I read as a child has stayed with me for decades and was all the information that I needed to reach my decision to not even remotely consider ingesting bleach as a viable option to combat the virus. When controversial topics hit mainstream media, one will find information and misinformation along with conflicting and opposing points of view. The aftermath of Trump's advice was covered internationally and from varied perspectives. An article on IndiaTVnews.com shares,

> "As per the official figures, the deaths due to household disinfectants in the months January and February 2020 increased by 5 percent and 17 percent respectively while the months of March and April have seen an astronomical rise of 93 percent and 121 percent."

A post from USnews.com states,

> "A new survey from the Centers for Disease Control and Prevention found that 4 percent of respondents consumed or gargled diluted bleach solutions, soapy water and other disinfectants in an effort to protect themselves from the coronavirus. Those people were among nearly 40 percent who reported using at least one method not recommended by the CDC in an attempt to reduce their chances of contracting the virus."

The Harvard Business Review weighed in with this,

> "While 4 percent may not seem like much, if this study sample was representative of the U.S. population, it would imply that roughly 12 million Americans engaged in these dangerous behaviors — an alarming figure indeed. But there may be reason to question that conclusion."

The article further detailed how the data could have some serious flaws and continues,

> "So how many Americans actually ingested bleach to ward off the coronavirus? We don't really know. We would need more research to reliably answer that question, but the fact that the percentage dropped from 4 percent to 0 percent after accounting for basic data quality issues suggests that the real number is most likely a lot lower than headlines would suggest."

My point here is not to tell you who or what to believe. My point is that we must examine multiple sources using logic as a lens to make an informed decision on not only what to do or believe but also understand our motivations in doing so.

As we endeavor to create a more equitable and just society, we must sharpen our consciousness for change. Acquiring an understanding of the history of race and racism in this country is a good place to start. That said, CRT deserves delving into with a learning mindset. It is the beginning of knowing why parts of American history have been hidden or omitted. As we look to CRT to unravel systemic and institutionalized racism in the United States, we must examine the narratives both for and against. I believe that far more Americans have learned of CRT from its critics rather than from the theorists themselves. Opponents are portraying CRT as anti-White and peddling a sense of personal guilt on other White Americans. They proclaim with certainty that CRT makes White children feel guilt that they should not feel about the atrocities endured by Black, Asian, and Indigenous people at the hands of their ancestors. I'm here to let you know that CRT guilt is an option that no one has to accept. No one is required to take CRT implications personally but should use them as an educational opportunity and learn the truth of American history and its role in racism that continues to this day. Create space for people to question whether they want to identify with prior generations and offer an enhanced perspective to remove oneself from it. We can educate ourselves on race and racism without baking in guilt. Most antagonists that I've encountered admit that they don't know what CRT is, believe that it's bad for our country, bad for our kids, and that they just don't like it. Those are really strong beliefs about something that one knows nothing about. If you are in this camp or just want to, in all fairness, learn more about it, it's sensible to start your CRT understanding by hearing what CR theorists say about themselves. Derrick Bell, Kimberlé Crenshaw, and Richard Delgado, all legal scholars at Harvard Law School, developed CRT as a body of thought in the 1980s. A post from the American Bar Association states,

> "CRT is not a diversity and inclusion 'training' but a practice of interrogating the role of race and racism in society that emerged in the legal academy and spread to other fields of scholarship. Crenshaw—who coined the term 'CRT'—notes that CRT is not a noun, but a verb. It cannot be confined to a static and narrow definition but is considered to be an evolving and malleable practice."

The article goes on to say that

> "CRT recognizes that racism is not a bygone relic of the past. Instead, it acknowledges that the legacy of slavery, segregation, and the imposition of second-class citizenship on Black Americans and other people of color continue to permeate the social fabric of this nation."

Our country was founded on the idea of equality. When we have systems and institutions embedded with inequality, there's going to be a struggle to reconcile the two. CRT not only provides a framework of thinking about race and racism but a way of questioning the status quo. The *New York Post* offers this perspective in an opinion piece by Christopher F. Rufo,

> **"Critical race theory draws heavily from black-nationalist ideology, such as that of the Black Panther Party, which came to fruition in California prisons in the 1960s. The new iteration of this ideology might have abandoned the militant rhetoric of the Panthers in favor of the therapeutic language of the school psychologist, but it nevertheless threatens to replicate the destructive features of prison-gang politics in the outside world. If American institutions succumb to this ideology, they can expect a brutal future: the suspension of individualism in favor of racial collectivism; a nihilistic, zero-sum vision of society; and endemic racial conflict as a baseline condition."**

Everyone has an opinion, and freedom of speech is our right. However, when it comes to building sustainable inclusion, it's essential to separate opinion from fact. When researching race and racism, we may be introduced to terms with which we are unfamiliar. We must take the time to investigate those terms, discern how they shape the message, and spin that through our logic. I never said that this education was going to be easy—nothing worthwhile ever is. As we are on an inclusion journey, equality is a front-seat passenger, and we must check incoming information for alignment with our commitment to inclusion.

Champions of inclusion examine points of view from varied sources and evaluate findings through a lens of commitment to culturally competent environments. After we've done our CRT research, we may find that we are not in agreement with its merits—and that's okay. At least you know now what you're disagreeing with. Disagreement with the facts does not render them invalid. Slavery, segregation, the fight for civil rights, and the genocide of Indigenous people are a part of this nation's history, regardless of how we feel about it and its effects. But we owe it to ourselves to decide for ourselves how and whether we will enact change and not let others decide for us.

Actions

Understand What CRT Is and What It Isn't

Knowledge is power. Acquire CRT facts for yourself in support of your inclusion journey. Decipher fact from fiction, expand your view of race and racism, and decide how to go about addressing inequality in a way that you're comfortable with. Here are a few CRT resources to get you started:

- **Book**: *Critical Race Theory: An Introduction,* by Richard Delgado, Jean Stefancic, et al.

- **Book**: *Faces at the Bottom of the Well: The Permanence of Racism,* by Derrick Bell and Michelle Alexander (2018)

- **Book**: *Critical Race Theory: The Key Writings That Formed the Movement,* by Kimberle Crenshaw, Neil Gotanda, Gary Peller, Kendall Thomas (1996)

- `Brittanica.com`: "Critical Race Theory": `www.britannica.com/topic/critical-race-theory/Basic-tenets-of-critical-race-theory`

- `WeForum.org`: "What is Critical Race Theory?": `www.weforum.org/agenda/2022/02/what-is-critical-race-theory`

Observe How Race and Racism Are Reflected In Your Workplace

With more and more organizations working to improve workplace diversity, racism continues to rear its ugly head. It can manifest in the form of racial slurs or jokes, decisions to hire/fire/promote based on race, or leadership's failure to address known incidents. Perhaps you're aware that a person of color was prohibited from attending a client meeting as they were deemed "not a good fit" for the client. Maybe there's a company policy banning certain hairstyles or headwear, which seems fair on the surface as it seemingly applies to everyone; however, only Blacks and Muslims are impacted by the policy. The effect of workplace racism is systemic and is a symptom of the systems in place. The *Business*

Insider post states that "Glassdoor surveyed 5,241 adults in the US, UK, France, and Germany. American workers were more likely to experience discrimination than all other countries." While most of us don't have the authority to institute company polices and processes, we can increase our awareness of racist acts and systems and inform management, human resources, or the Equal Employment Opportunity Commission (EEOC) or use other organizational tools like an anonymous tip line to call out discriminatory practices to effect change. For more details on saying something, review Activity 26, "If You See Something, Say Something."

Hold Yourself Accountable

The journey to inclusion isn't easy and is ever-changing. It's important to be aware of the unique issues that people of color face in the work-place and be mindful not to personally perpetuate them. Mistakes will be made and rather than claim unintended harm and dismiss the accu-sation, hold yourself accountable. Acknowledge it, listen to the person you've offended, apologize, and use it as a learning opportunity to do better in the future.

Action Accelerators

- **BusinessInsider.com**: "42% of US employees have experienced or seen racism at work. It's the latest example of how diversity efforts are falling short, especially in America," by Allana Akhtar: www.businessinsider.com/glassdoor-42-of-us-employees-have-witnessed-or-experienced-racism-2019-10

- **TheEWgroup.com**: "Racism at Work: How to Stop Discrimination in the Workplace," by Jane Farrell: theewgroup.com/us/blog/racism-work-stop-discrimination-workplace

- **ISB.idaho.gov**: "Critical Race Theory and Workplace Diversity Efforts," by Bobbi K. Dominick: isb.idaho.gov/blog/critical-race-theory-and-workplace-diversity-efforts

Sources Cited

IndiaTVnews.Com. US deaths due to poisoning rise after Trump endorses bleach, disinfectant as COVID-19 treatment, www

.indiatvnews.com/news/world/us-coronavirus-deaths-by-bleach-disinfectant-injection-major-rise-trump-covid-19-treatment-616708

Cecelia Smith-Schoenwalder. CDC: Some People Did Take Bleach to Protect from Coronavirus, USnews.com, June 5, 2020, www.usnews.com/news/health-news/articles/2020-06-05/cdc-some-people-did-take-bleach-to-protect-from-coronavirus

Rachel Hartman. Did 4% of Americans Really Drink Bleach Last Year? HBR.org, April 20, 2021, hbr.org/2021/04/did-4-of-americans-really-drink-bleach-last-year

CDC.gov: Cleaning and Disinfectant Chemical Exposures and Temporal Associations with COVID-19, National Poison Data System, United States, January 1, 2020–March 31, 2020, www.cdc.gov/mmwr/volumes/69/wr/mm6916e1.htm?s_cid=mm6916e1_x

NYPost.com: Critical race theory is about to segregate America like an open-air prison yard, nypost.com/2022/02/04/critical-race-theory-is-about-to-segregate-america-like-an-open-air-prison-yard

Janel George. A Lesson on Critical Race Theory, AmericanBar.org, January 11, 2021, www.americanbar.org/groups/crsj/publications/human_rights_magazine_home/civil-rights-reimagining-policing/a-lesson-on-critical-race-theory

TELLING IT LIKE IT IS. . .TO GET WHERE WE NEED TO GO

Racism Leaves No Room for Equity

Charles (not his real name) began his career in the manufacturing industry as a general laborer at 20 years of age. He spent years developing professional skills, investing time to learn the tools of trade, and stepping in to be cross-trained for other roles in hopes of achieving a senior leadership position and leading an entire facility or region someday. He was determined to be the success story of starting at the bottom and working his way up to incredible heights. Seemingly, as long as he was at a blue-collar level and being paid hourly, advancement and annual raises were achievable with hard work. Within four years he advanced to a supervisor role and eventually earned an office position as a scheduler. He went from scheduling the runs on various machines to scheduling the entire facility and purchasing tens of millions of dollars of roll stock annually. He kept his head down, became known for being a hard worker, and never called in sick. He shared his knowledge by mentoring his younger and less experienced colleagues. By all accounts, he was a model employee. During his 10th year at a large family-owned box manufacturer, he finally earned a salaried position as a purchasing manager and was given an office. In his mind, he was on his way up the corporate ladder at 35 years of age. He was the only African American with a manager's title and the only person of color with an office.

The honeymoon quickly ended when a new regime of White leaders came onboard and yanked his plum position in favor of one of their own. Without warning or explanation, they demoted him back to the manufacturing floor where he began his career a decade earlier. The subsequent discussion was short and bitter: take the offer or resign. Charles had a family to feed and a mortgage, so he swallowed his pride and endured the embarrassment and humiliation. Those feelings quickly turned to anger and insult when he was asked to train his replacement. How does one process such a thing? The audacity to take a position from someone who successfully held the role for almost a year only to be replaced with someone less qualified was unfathomable. He suspected racism as the root cause of the decision and saw it validated without question when his replacement knew nothing about purchasing, while Charles had to earn the role by proving himself capable. He later learned from the CEO's admin, a woman whom he had developed a trusting relationship with over the years, that the original plan was to fire him. However, one of the owners, having remembered Charles's value in the plant, intervened and recommended that he return to his former duties, thus saving his job. Perhaps the owner thought that he was doing Charles a favor, but it certainly did not feel like a favor. Charles no longer worked the normal business hours of 8 a.m. to 5 p.m. He was relegated to the third shift—midnight to 8 a.m.—and forced to work weekends regularly. Charles wondered about the owner's true motivation for intervening. Was he sincerely trying to save the job of a dedicated employee, or was he acting in self-interest by keeping the skills in-house on a shift that was in shambles?

Returning to his former role did not go smoothly. Charles expected to resume his former duties and planned to perform his best, as he had always done, while working on an exit strategy. He was determined to be a class act, no matter what. In the interim of Charles's role in purchasing, a new position had been created in the plant where he now reported to someone new, Dave. Dave was the plant superintendent who was also White. Charles wondered why he was not considered for that role. He was certainly qualified. It didn't take Charles long to realize that he was an unwelcomed member to Dave's team. Suddenly Charles could never get anything right and was held accountable for things beyond his control. Week after week, Dave reprimanded Charles for something. Prior to reporting to Dave, he had never been written up for misconduct or placed on probation and always received good performance reviews. Over time, people resigned or were fired under Dave's leadership. Rather than replace them, Charles was frequently scheduled to cover additional shifts and often survived on five hours of sleep. Charles was living a nightmare. He was helpless in fighting what he called "city hall." The problems were systemic. The schedule made it impossible to seek employment elsewhere as there was barely time to eat and sleep, let alone time for identifying jobs for which to apply. The idea of making time for interviewing especially in a perpetual state of exhaustion just seemed pointless. He joked with his wife that he would probably fall asleep during an interview and would be lucky if he could stay awake long enough to complete an application. Between the exhaustion and the stress, Charles' health began to deteriorate. While Charles' family and friends recognized that he was wasting away and feared for his life, Dave complimented him on the weight loss and told him how much better he now looked. The racism could no longer be ignored or tolerated. Charles refused to sacrifice his health for any reason, but the idea of doing so for the betterment of the company's bottom line repulsed him. Racism-induced stress is an overwhelming cocktail that no one should be expected to drink. The circumstances surrounding the demotion, combined with Dave's mistreatment and utter disregard for the human condition, assured Charles that he would never be treated equitably in that organization and made it easy to resign before landing a new job. The torment he experienced working in the industry left a harsh taste. Charles left the industry and never looked back.

Go Beyond Performative
Gestures

"Actions speak louder than words."

—**Proverb**

***

When what we say and do in public is incongruent with what we say and do in private, it is considered performative and hypocritical. Social movements like #MeToo, Black Lives Matter, Gay Rights, or Stop Asian Hate tend to reduce us to performers and hypocrites when we want to be perceived as caring about justice and equality for marginalized groups. But wouldn't it be easier to just be authentic than be discovered a fraud? For many individuals and organizations alike, the answer is absolutely not! The pressure may be too intense. Perhaps there's guilt in not caring or we just want to jump on the bandwagon and get congratulated for being there. The last thing we want is for others to believe or suspect us to be racist, sexist, homophobic, transphobic, or xenophobic, especially when it takes minimal effort and no commitment to equality and justice to showcase a Pride mug or other symbolic gesture on our desks to assuage the guilt and pressure. In doing so, we get to feel better about ourselves and reassure our marginalized co-workers or staff that we're one of the good ones. In essence, we're saying "I am on your side. I support you," without having uttered a word. The truth is that symbolic gestures never equate to action (See figure 37.1). They are just words and symbols masquerading as support, and eventually people will discover the truth. How will one respond if ever called on to back up those signals of support? A post by PsychologicalScience.org shares,

---

"We're averse to hypocrites because their disavowal of bad behavior sends a false signal, misleading us into thinking they're virtuous when they're not, according to findings in *Psychological Science*, a journal of the Association for Psychological Science. The research shows that people dislike hypocrites more than those who openly admit to engaging in a behavior that they disapprove of."

The post continues,

"Intuitively, it seems that we might dislike hypocrites because their word is inconsistent with their behavior, because they lack the self-control to behave according to their own morals, or because they deliberately engage in behaviors that they know to be morally wrong. All of these explanations seem plausible, but the new findings suggest that it's the misrepresentation of their moral character that really raises our ire."

Figure 37.1: Symbolism is not enough.

That said, rather than waste so much energy in pretending to stand for equal rights, opting to do nothing may be a close second. That way, we're not standing for or against, and we get to enjoy the comfort of the neutral zone. Spoiler alert: doing nothing is doing something—it condones and perpetuates the very systems and behaviors that need to be changed (See figure 37.2).

Figure 37.2: Become part of the solution.

People label themselves an ally in acts of performative allyship and believe that they can skip doing the actual work. I've never seen where the label matches the sentiment in absence of the work. When we're not using gender-neutral language in mixed groups, recognizing and calling out microaggressions, or engaging in conversations on polarizing topics and so on, we are not doing the work but engaging in performative and symbolic gestures that miss the mark by a long shot. Folks may argue that the symbolism serves as a reminder of their allegiance and a call to be prepared when they see something or when asked for help. That's like preparing for an exam by merely looking at the study materials—totally ineffective. Individuals from marginalized groups are well aware of the long history of the dominant culture calling themselves allies and yet have nothing to show for their alleged support; or worse, the supposed allies are continuing to inflict harm. Millions donned a safety pin as a symbol of support for underrepresented groups amid the Donald Trump campaign and subsequent election to represent allyship. The campaign message was a direct threat to the health, safety, and well-being of Black and Brown Americans, and wearing the pin signaled "safety" and support for people of color as well as a willingness to stand up if and when needed. As of this writing in 2022, I don't see the pins anymore. What symbol should marginalized groups now look for? Unless you're doing the work, you can't be identified and won't be considered an ally. The election of Joe Biden did not eradicate inequitable systems and institutions, and the threat persists as long as there is racism and prejudice. There's still much to be done. Moving past performative allyship means asking, "How can I show that I stand for justice and equality in an authentic, impactful, and accountable way?" Actions speak louder than words, and words and actions must be in sync, especially in these divisive times. Wearing the "Black Lives Matter" or "Stop Asian Hate" T-shirt, displaying a Pride flag at your workstation, or posting your sentiments about injustice on social media are all great ways to raise awareness, but being aware of a problem won't solve it. Awareness is only the beginning. Social movements exist for a reason: to help disrupt systems of oppression, discrimination, and prejudice. Effective allies must first acknowledge that inequalities exist and the role they play in perpetuating them and then find the courage to shift from perpetuating the problem to eradicating the problem. We must come to terms with the fact that each of us is in a position to drive change. Everybody can do something in one way or another to varying degrees. The more power and privilege one has, the better the impact to be made. While some may be in a position to lobby for gay rights, others are able to provide financial support from a few dollars to thousands of dollars to nonprofit organizations that support gay rights. While some of us are in a position to hire and promote deserving and qualified individuals from historically excluded groups, others can mentor or sponsor them, while still others can connect, recommend, and write referrals for jobs, scholarships, and more. Champions of inclusion endeavor toward tangible and lasting change that improves lives and ultimately systems and culture. We acknowledge our privilege and use it for the greater good, rather than personal gain. True allyship is a verb. It's intentional, proactive action with consistency. Starting with individual actions helps in learning the issues and honing skills. I find this to be the comfort zone for many. Others may be visionaries and innovators who can institute collective action and cultural shifts. Changes in systems and institutions can accelerate as members of the dominant culture work to stimulate rather than stymie transformation, especially in those areas that members of the minority culture can't access due to lack of power and privilege. It's essential to face the fear in making the first step and taking it anyway. The fear dissipates over time. Active allyship is an ongoing journey toward diversity, equity, inclusion, belonging, and justice for marginalized communities, thus making the workplace better for all, as well as society as a whole.

Actions

Be Real with Yourself

Our hypocrisy enables us to fool folks for a short time, but we can never fool ourselves. Performative allyship is not allyship but a fragile façade. No one truly benefits from hashtags, slogans, and symbolic memorabilia. We must decide whether we are in or out of active allyship. There is no neutral. Just realize that being out means we are condoning or, worse, perpetuating systems of oppression and discrimination.

Learn to Recognize Personal Biases and Racist Tendencies

I know, I know. Everyone is biased, and no one admits to being racist. However, racism and bias exist every day both inside and outside the workplace. Our culture is embedded with discriminatory practices against anyone who is *not* cisgendered, white, male, heterosexual, and able bodied; thus, all of us are capable of racist or prejudicial behavior. Many of us are oblivious to it, especially when we are the culprits. Nevertheless, everyone has a role in creating safe and inclusive environments, and as allies advocating and fighting for equity, the work begins on a personal level. Review Activities 1, 2, 3, 12, and 13 for guidance. The greatest hypocrites are those who call out others for the very actions they commit themselves. Checking for our biases and intervening when racism or oppression occurs in the workplace requires educating ourselves about ourselves. Further, learning how racism and inequality shows up is essential if we are to be effective. We must realize how our beliefs and tendencies impact not only people of color but also where people intersect, i.e., an Asian, bisexual female. In this intersection of identities, an Asian male would naturally advocate for other Asians but may discriminate against a queer, Asian female.

Move from a Place of Comfort to Growth

Performative allyship is a place of comfort. There is no risk, no commitment, and no action—only words and symbolic gestures. However, I'm willing to bet that since you made it this far in the book, you are well-intentioned, open-minded, courageous, ready to learn, and socially conscious. We must now embrace the discomfort or inertia that we've deflected for so long so that growth can begin. This may require moving past guilt and fragility for most, while for others it's about learning to leverage the privilege you have. Everyone has some level of privilege that can help someone else. Moving from comfort to growth won't be linear; there will be setbacks, and that's okay as long as we get up and get back out there. Exposure to ambiguity and volatility in this space scares the heck out of folks, so anticipate it, build resilience, and eventually, you'll thrive in it. It's how allies become effective.

Review allyship steps and considerations to move from performative to active:

- ▪ ncaaorg.s3.amazonaws.com/ncaa/programs/a4/LD_A4Allyship.pdf

- ▪ www.ubs.com/global/en/assetmanagement/insights/thematic-viewpoints/sustainable-impact-investing/articles/what-does-it-take-to-be-an-ally/_jcr_content/mainpar/top levelgrid/col1/innergrid_1475432314/xcol2/textimage_377377446.1742344515.file/PS9jb250ZW50L2RhbS9hc3NldHMvYW0vZ2xvYmFsL2luc21naHRzL3dlYmluYXIvaW5nL2FsbHlzaGlwLnBuZw==/allyship.pngf

Action Accelerators

- **JenniferBrownSpeaks:** Podcast: E113, "Pain and Possibility, How Dr. Maysa Akbar is Charting What's Beyond Allyship": `willtochange.libsyn.com/e113-pain-and-possibility-how-dr-maysa-akbar-is-charting-whats-beyond-allyship`

- **Ginger.com:** "Beginner's Mindset: How to Learn & Grow as an Ally" by Elissa Burdick: `www.ginger.com/blog/beginners-mindset-how-to-learn-grow-as-an-ally`

- **YouTube.com:** *Black lives don't need performative allies*, Déjà Rollins, TEDxUAMonticello: `www.youtube.com/watch?v=U5DU0fqF9Uo`

- **YouTube.com:** *Listen, Learn, & Speak Up: Allyship & Activism*, Chiara Lea, TEDxBrentwoodCollegeSchool: `www.youtube.com/watch?v=ec7aIpo9yJI`

- **YouTube.com:** *A guide to lifelong allyship*, Catherine Hernandez, TEDxToronto: `www.youtube.com/watch?v=g3D-5-2EqHI`

Sources Cited

PsychologicalScience.org. "We Dislike Hypocrites Because They Deceive Us," January 30, 2017, `www.psychologicalscience.org/news/releases/we-dislike-hypocrites-because-they-deceive-us.html`

38

Reconsider the Golden Rule

"Cultural integration doesn't happen by you boasting about your culture; it happens by you coming forward enthusiastically to learn about another culture."

—**Abhijit Naskar**

***

Most everyone is familiar with the golden rule: "Do unto others as you would have them do unto you" (Luke 6:31). To put it another way by most non-Christians, "Treat people the way you want to be treated." The message is consistent across many cultures and religions including Buddhism ("Treat not others in ways that you yourself would find hurtful" [Udana-Varga 5,1]) and Islam ("Do not let hatred of others lead you away from justice, but adhere to justice, for that is closer to awareness of God" [Qur'an 5:8]). In both religious and secular realms, it's considered sage advice that many can agree with. We may have been taught this adage at an early age when our parents spoke of the importance of being a friend in order to make a friend or when we called our siblings selfish for not sharing as we have shared. The lessons learned along the way contributed to our sense of fairness. Today, they are words that we continue to live by as adults. By the time we enter the workplace, we are certain of how we want to be treated and strive to treat others with the same respect. It seems like the perfect default when it comes to building inclusive and equitable workplace cultures. We simply treat everyone in accordance with our preferences in hopes that they'll reciprocate in kind. The practice requires little to no thought, and it seems fair. We say to ourselves, "I don't mind being addressed as *guys*" even though I am a woman. So, it's okay when I address other women the same way; or "I enjoy a good joke to break the monotony of work, so I'll share the one I just heard about the Black man at the bar." While the actions seem innocent on the surface with no ill-intent, we haven't paused to consider how others may feel or react. All women don't want to be addressed with "Hey, guys," and everyone won't find the joke about a Black man in a bar humorous, especially if it's perceived as racist. Our assumptions make it difficult to understand when we unintentionally offend someone when we apply the golden rule. The premise has major flaws as we interact with individuals from different cultures, backgrounds, and experiences than our own. In fact, some norms that are perfectly acceptable in one culture may be offensive or harmful in others. White women may feel admired when a colleague complements their hairstyle and gives it a soft touch, while Black women may feel violated by the intrusion to personal space, especially when done by non-Black colleagues. Members of the dominant culture may feel praised when told that they are very articulate, while Black people may feel invalidated by the comment. It's plausible that the failure of White people and people of color to understand one another's perspective is perpetuated by the golden rule. As an individual from a specific culture, background, lived experience, and personal set of standards, we can know only how *we* want to be treated based on that construct. This is why cultural competence is so important as well as understanding that any principle will have its limitations when we try to apply it across the board to all co-workers and business scenarios all the time. People become disillusioned into thinking that just because something feels good, sounds good, and is accepted by them personally, that it must be universal. Assuming all co-workers should be treated in the same way that we ourselves want to be treated imposes our personal preferences and values on those with whom we interact.

Abandoning the golden rule for the more useful platinum rule serves as a much better foundation for creating more inclusive interactions. I was first introduced to the platinum rule, "Treat others the way they would like to be treated" more than two decades ago when I read the book *The Platinum Rule: Do Unto Others as They'd Like Done Unto Them*, by Tony Alessandra and Michael J. O'Connor. Later, in the book *The Art of People* by Dave Kerpen, I gained a new perspective about the golden rule, which in short was that what I want does not mean that it is also what others want. Discovering what others want and how they want to be treated means making it our business to find out through building trusting relationships, reading body language, asking questions, and observing our surroundings. We must make space for co-workers to communicate how they want to be treated. Inclusive behaviors demand a learning mindset and openness to understanding the unique experiences of others. Champions of inclusion acknowledge that everyone is different and has unique preferences. We endeavor to create meaningful connections to better understand how others want to be respected and valued.

Actions

Create a Baseline

Consider which personal preferences may be widely acceptable to begin practicing treating people the way *they* want to be treated. Create a "how I like to be treated list" of at least 40 ways you like to be treated in the workplace, i.e., "I like people to look me in the eye when we talk." Examine whether these preferences are based on your culture, values, and upbringing and which may be common to other cultures or are simply basic human needs. Use these as a baseline and avoid all others until you get to know individual preferences.

Honor the Signals

I stated in a previous chapter that actions speak louder than words, and body language speaks volumes. When learning the unique preferences of others, focus on body language and other reactions that signal discomfort or misalignment with your personal preferences. This may come across as being soft spoken versus your norm of speaking loudly, inferences to get to the point, or maintaining a comfortable distance. I once had a boss who had a habit of hugging. When I stepped back and extended a handshake, she became offended and never let me forget it. I was offended as well, and my level of comfort was never considered. Be cognizant that behaviors don't apply across cultures.

Review Activity 5, "Experience Other Cultures," for guidance on building cultural competence.

Re-evaluate an Encounter Gone Wrong

No one is perfect. We are all guilty of miscommunications, misunderstandings, and misinterpretations. Reflect on the last time you found yourself struggling to understand and be understood. Sort out the dynamic of the encounter by answering the following questions:

- Was your interaction focused more on your needs or that of the other individual?
- In what ways was the other individual different from you?
- How were you considerate and responsive to the other's needs, feelings, capabilities, etc.?
- How were your actions received?
- If you had a chance to do it again, what would you do differently?

Action Accelerators

- **YouTube.com:** *The Trouble With The Golden Rule*, Brendan Schulz, TEDxYorkU: www.youtube.com/watch?v=4-BYhAcabZk&t=320s
- **YouTube.com:** *Verna Myers on Relationship Building*, Representation Matters, Salesforce: www.youtube.com/watch?v=LywU1BFpxr4
- **Stitcher.com:** Jennifer Brown Podcast, *The Will to Change*, Minisode #13: "Beyond the Choir-Bridge-Building for Change": www.stitcher.com/show/the-will-to-change-uncovering-true-stories-of-diversity-and/episode/minisode-13-beyond-the-choir-bridge-building-for-change-54905844
- **HBR.org:** "3 Small Ways to Be a More Inclusive Colleague," by Juliet Bourke: hbr.org/2021/12/3-small-ways-to-be-a-more-inclusive-colleague

Sources Cited

Bessel.org. www.bessel.org/golden.htm

Eliminate Double Standards

"The cost of liberty is less than the price of repression."

—W. E. B. Du Bois

***

Those of us who grew up with at least one sibling may have experienced times of partaking in the exact behavior as our brother or sister only to be chastised for it, while our sibling continued to get away with the behavior without consequence. Daughters may have been given earlier curfews than sons. The eldest sister may have been prohibited from wearing makeup until 16 years of age, but when the younger sister turned 13, mom gifted her a makeup palette. Perhaps a low report card grade earns one sibling a loss of phone privileges for a month while the other is advised to "do better next time." Growing up, my love of designer clothing and shopping was shared by both my sister and me. However, my parents were very judgmental of my spending habits but looked the other way when she splurged as if it were no big deal. Children feel helpless when they are victimized by double standards whether it occurs at home or school. How does one discern fairness and expectations when rules are not consistent? For kids, it's confusing at best and painful at worst when adults unintentionally send signals that double standards are the norm. When questioned by the young and impressionable mind, adults defend their decisions by providing a flimsy justification, denying the accusation, or becoming hostile for being called out. Regardless of their logic in imposing the double standard, it still felt unfair. The workplace is also littered with double standards. `Dictionary.Cambridge.org` offers two definitions:

- The habit of treating one group differently than another when both groups should be treated the same
- A rule or standard of good behavior that, unfairly, some people are expected to follow or achieve but other people are not

This unfair application typically leads to the repression of a group or individual. While working as a store manager for a national retailer, I was supposed to adhere to a strict dress code defining my attire from head to toe. For the most part, peers followed the code with little deviation. Over the months I noticed colleagues enhancing the otherwise mundane look with personal accents like swapping black shoelaces for white ones, or the black belt with black buckle for a black belt with a gold buckle. Not being a fan of the dress code and following the lead of my more tenured peers, I made a few subtle substitutes of my own. I actually enjoyed the work more when I expressed my individuality. One afternoon at the beginning of my shift, my boss beckoned me to enter his office and to close the door. He presented me with a document informing me of my misconduct in not adhering to the dress code. After he confirmed that I was in fact aware of, and understood, the dress code, he insisted that I sign the document. When I refused, citing that I was simply following the lead of his other direct reports and assumed the small adjustments were acceptable, he claimed to never have noticed their deviations and informed me that whether I agreed, disagreed, signed, or didn't sign, the document would become part of my personnel file. I could not convince him to see the circumstances from my perspective and felt singled out. As soon as the encounter ended, I contacted a few peers whom I could trust to inquire whether they had the same experience. As a member on a racially and gender-diverse team of 12 who reported to an African American male, I could not make sense of being singled out. The men were surprised that it happened. Of the two women, one shared that she never deviated from the code and the other, who happened to be White, shared that she had been given only a verbal warning. Aha! So, my boss had noticed, at least once, but chose not to give me a verbal warning and, for some reason, opted for something harsher. His actions convinced me that he was sexist at the very least. I'd never fancied his style of leadership, and on that day he lost his credibility and acceptance as a trustworthy leader. No one can work to their full potential under a leader who employs double standards in decision-making—unless, of course, they are a beneficiary. Leaders are not the only culprits. Peers and colleagues are guilty as well when they manipulate, distort, and omit details or use selective memory to fit an agenda favoring one group or individual over another. Double standards in the workplace are inevitable as most of us have either had to navigate them or employed them from a young age; and while everyone agrees that they are harmful, they are practiced nonetheless and sometimes at an unconscious level.

As the workforce continues to become more diverse, the perception of intelligence, professionalism, and leadership qualities continue to default to White males in many industries as biases and prejudice prevail. Conversely, women and members of marginalized groups tend to be undervalued and held to different or higher standards than their White, cisgendered, heterosexual, male counterparts and can be judged or criticized for exhibiting identical behaviors. In many cases, the difference is based on cultural upbringing. For example, women are perceived as nurturers, and when they opt to exhibit more masculine tendencies to be viewed as a credible leader, they are deemed rude or abrasive. It's as if women have to be one or the other but can't be both. How people regard or perceive one another makes a big difference in relationship building, inclusion, and belonging. When a White male is assertive, he's the boss. When a woman is assertive, she's bossy. When a White male exhibits strong feelings about a topic, he's passionate. When a Black male expresses the same sentiment, he's angry and advised to adjust his tone. Double standards affect our perception of competence. When a man's hair starts to gray, he is considered more distinguished and credible, while women are viewed as aging and thus less competent. Women never seem to be the right age. We are either too young to be taken seriously or too old. Many believe that the more attractive the woman, the less intelligent she must be, and she is thought of as a bimbo. Attractive men perceived as less intelligent enjoy a jock or stud status without question of their ability to do their job. If a woman is both attractive and successful, she must have slept her way to the top. When a man is attractive and successful, no one credits his success to his looks but more his talent. A man's neutral facial expression demonstrates strength, when women are advised to smile more to erase the "resting bitch face" expression.

Folks are quick to agree that Black, Brown, and LGBTQ individuals deserve equal rights, equal access, and equal opportunity, yet those are all denied when they are held to different standards. While strides have been made with diversity in the workplace, inclusion is ultimately hindered as the goalpost shifts when double standards are at play. The next time you catch yourself referring to a female colleague as bossy or questioning how she achieved her position or advising a co-worker of color to tone down their level of intensity, that's cause to pause and question belief systems as gender and racial stereotypes remain firmly entrenched. We must also consider the personal impact of mistrust and resentment when our co-workers take note. Champions of inclusion work by a standard of trust over double standards and maintain a mindset of fairness to all even when it's not convenient. When considering common double standards in the workplace (see Figure 39.1), it's worth pointing out that women of color experience double standards based on skin color as well as gender.

WOMEN	MEN
Leadership = Competent *or* likeable, seldom both	Competence and likeability coexist in leadership roles
Attractive + successful = slept their way to success	Attractive + successful = Talented
Comes across as abrupt, intense	Comes across as decisive, in control
Assertive, driven = bossy	Assertive, driven = ambitious, leader
Perform menial tasks = taken for granted	Perform menial tasks = acknowledged, appreciated
Inebriated at the office party = reckless, harshly criticized	Inebriated at the office party = unspoken or glossed over
Speaking in vulgar four-letter words = unprofessional, unladylike	Speaking in vulgar four-letter words = brutal honesty
Demonstrating negative emotion = emotional	Demonstrating negative emotion = pressured
Kind, friendly = flirtatious	Kind, friendly = kind, friendly
Consensual affair with older, wealthy boss = gold-digger and stigmatized	Consensual affair with older, wealthy boss = stud, accepted and congratulated
Consensual affair with younger, attractive associate = cougar, sugar mama and stigmatized	Consensual affair with younger, attractive associate = kudos and acceptance

PEOPLE OF COLOR	WHITE PEOPLE
Characterized as angry or aggressive	Characterized as passionate
Direct, outspoken = offensive, threatening	Direct, outspoken = candid, forthcoming
Refusal to perform nonjob related menial tasks = uppity, not a team player	Refusal to perform nonjob related menial tasks = knows their value
Asking for help = unqualified	Asking for help = seeking guidance
Working late = neglecting duties during regular hours	Working late = hard working
Fatal mistake = incompetent	Fatal mistake = lapse of judgment
Consistent overachiever = met with surprise, must be cheating, work more scrutinized	Consistent overachiever = no surprise reaction, attributed to strong work ethic, empowered to improve
Bending the rules = unethical, dishonest	Bending the rules = playing to win
Sterling and impeccable work before career advancement	Good and good enough work before career advancement

Figure 39.1: Common double standards in the workplace

Actions

Hold the Privileged to the Same Standard That We Hold the Marginalized

Lower or different expectations are the norm for marginalized groups based on racism, sexism, and stereotypical beliefs and impact how competence is perceived. When expectations are exceeded, they are met with disbelief, and achievement is considered out of the ordinary. Why? Because marginalized groups are not expected to succeed, nor were systems designed in support of our success. A shift in thinking toward equal expectations of all places everyone on the same level regardless of age, race, gender, ability, or sexual orientation and allows us to view things objectively rather than through preconceived notions. Refer to Activity 12, "Strive to See the Whole Person," for guidance. We can mitigate, and ideally one day eradicate, the need for double standards to justify decisions and behaviors when the privileged are held to the same standard as the marginalized.

Become Part of the Solution

Double standards are pervasive and will continue to impact the workplace when we stay silent. When we see something, we must say something. Refer to Activity 26, "If You See Something, Say Something," for guidance.

Always Choose to Do the Right Thing

Employing double standards is easy and convenient. They can support goals, agendas, and the desire to be right at any cost. We come up with excuses to rationalize double standards and think nothing of it, not even the fact that it harms us in the long run by eroding trust. Let's endeavor to think more about it as well as its impact. In the process, we become better people and create better workplaces and societies while demonstrating our commitment to integrity when we always choose to do what's right, not what's convenient.

Action Accelerators

- **SHRM.org**: "Barriers for Black Professionals," by Lisa Rabasca Roepe: www.shrm.org/hr-today/news/all-things-work/pages/racism-corporate-america.aspx

- **NPR.org**: How to Survive in a Mostly White Workplace: Tips for Marginalized Employees," by Andee Tagle and Anjuli Sastry Krbechek: www.npr.org/2020/09/10/911464100/how-to-survive-in-a-mostly-white-workplace-tips-for-marginalized-employees

- **Forbes.com**: "Not Very Likeable: Here Is How Bias Is Affecting Women Leaders": www.forbes.com/sites/pragyaagarwaleurope/2018/10/23/not-very-likeable-here-is-how-bias-is-affecting-women-leaders/?sh=2e0d03ce295f

- **YouTube.com**: *Girl boss—Navigating the double-standard of gender + power*: www.youtube.com/watch?v=Ff5ieaomk3c&t=17s

- **HBR.org**: "How Female Leaders Should Handle Double-Standards," by Herminia Ibarra: hbr.org/2013/02/how-female-leaders-should-handle-double-standards

Remove the (Color) Blinders

"The shift may, in fact, come as something of a relief, as it moves our collective focus away from a wholly unrealistic goal to one that is within anyone's reach right now. After all, to aspire to colorblindness is to aspire to a state of being in which you are not capable of seeing racial difference—a practical impossibility for most of us. The shift also invites a more optimistic view of human capacity. The colorblindness ideal is premised on the notion that we, as a society, can never be trusted to see race and treat each other fairly or with genuine compassion. A commitment to color consciousness, by contrast, places faith in our capacity as humans to show care and concern for others, even as we are fully cognizant of race and possible racial differences."

—Michelle Alexander

***

I've met people throughout the years who experience *actual* color blindness. My first discovery was during lunch when a colleague asked me the color of a banana he wanted to purchase from the cafeteria. He wanted to make sure that it was more yellow than green. I immediately assumed that he was kidding until he told me that he was color-blind. I still thought that he was joking so he shared how his wife must coordinate his wardrobe to ensure that everything matched, just so he could get to work in the morning. At that point, I was convinced. Thankfully, we had a trusting relationship, so no damage was done as I expressed my disbelief. The fact that I didn't believe his color blindness, didn't make it any less real. The Cleveland Clinic describes this as:

---

color vision deficiency (CVD)—a condition where you don't see colors in the traditional way. This can happen if certain cells known as photoreceptors, or more specifically cones, in your eyes are missing or not working correctly. These cones typically allow you to see each color on the rainbow. If you have color blindness, you might not see each of these colors.

When we hear the term *color blindness* in the workplace, seldom are we referring to someone's medical diagnosis, but rather an attempt to not see race. Researchers Nellie Tran and Susan Paterson describe racial color blindness as "a belief system that some individuals hold suggesting that race no longer matters and that the Unites States represents a meritocratic society where hard work is the sole determinant of success." The attempt to not see race does not render it invisible. In principle, racial color blindness seems harmless or even the perfect solution to racial injustice and inequality. When folks believe that race is truly invisible or goes unnoticed, it stands to reason that racist thoughts and behaviors can't happen—at least on an individual level. The pretense that one does not notice color assumes that everyone must be the same and share the same experiences. If one has never experienced racism personally or felt that their race has never negatively impacted them, they can project that experience onto others and therefore assume that it must not exist. Experience tells people of color that nothing could be further from the truth. Those endorsing color blindness hold a point of view that denies racial dynamics within society, that race is not a contributing factor to an individual's lived experiences, and that racism is a thing of the past. After all, slavery has ended, school systems are more integrated, it's no longer illegal for one to marry outside their race, and we elected a Black president twice. To validate this point of view we can even point to the success of prominent people of color like Ellen Ochoa, the first Latina astronaut; Black financier Ken Chenault; and Illinois senator Tammy Duckworth of the AAPI community and double amputee. This level of change did not happen because of color blindness but because of decades of hard fought-and-won battles by historically marginalized and oppressed people. Choosing to ignore color completely negates racial injustice and the fact that race has been and continues to represent struggle, division, and subjugation for Black and Brown people. White people are able to comfortably subscribe to color blindness because they are usually unaware of how race affects people of color in the workplace and in American society as a whole. The idea of a color-blind society, though well-intentioned, demonstrates less appreciation for diversity and falls short of fostering cultures of inclusion and belonging. Color is not invisible, no matter how blind to it one claims to be, and it continuously shows up in overt and covert ways.

The belief that race is not taken into account or even noticed for that matter as it relates to opportunity makes it impossible to justify the fact that 466 CEOs on the Fortune 500 List are White, according to a 2020 report by Statista.com. The report further explains:

> **Black, Asian, and Hispanic executives face more obstacles when navigating the leadership pipeline of various Fortune 500 companies. Overall, surveys show that minority employees at these companies see obstacles like less opportunities for advancement, smaller rates of retention and tougher recruitment barriers.**

The data shows that color is seen through a crystal-clear lens and that bias toward race persists in the midst of colorblindness. When I encounter a Black, Asian, or LatinX CEO, for example, I can rest assured that their ascension to the top involved overcoming more difficulty and adversity than a White CEO would have encountered. It's important for the dominant culture to truly understand that people of color walk through the world differently than they do and it's by design. We are expected to conform and assimilate to their norms to build relationships and have a chance of success in the workplace. It's not the other way around where the dominant culture must get to know us to achieve any level of success. No conformity or assimilation is required on their part. While people of color tend to not be color blind, we are in fact deeply impacted by its ideology.

In conversations with my White friends and colleagues on the topic of race, some share that they don't see race and that they treat everyone the same. They were raised to not see color and declare that we are all members of the human race and that's all that matters. They purport that they don't care whether one is Black, Brown, and jokingly red, yellow, or purple—it's all the same to them. I know that they mean no harm.

What they don't realize is that those well-meaning words erase me, my identity, my culture, and my heritage—all the parts of me that are sitting alongside them. If one can't see my color, then they can't possibly see me. Many who espouse color blindness tend to regain their sight when they feel that they've been treated unfairly. Skin color is top of mind when lamenting that they were passed over for promotion in favor of a deserving and qualified Black or Brown person. They feel that reverse racism was at play. By contrast, when they hear people of color lament about being passed over for promotion in favor of a deserving and qualified or lesser qualified White colleague, they feel that that individual was "the best person for the job" and race was not a factor. Hard work alone is no assurance of success for people of color and even more so where identities intersect to include LGBTQ individuals or persons with disabilities. A point raised in Activity 32, "Leverage Your Privilege," is that the workforce in the United States is comprised of 78 percent White people, while the 2020 census data reports that White people represent 57 percent of the total U.S. population. (That number would decrease if it reflected White people who are no longer in the workforce.) These statistics make a compelling case for race as a barrier to success and show that people are far less color blind than they claim to be. How will people of color in the United States ever have an equal shot at achieving the American Dream of success, prosperity, and upward social mobility through hard work when we are systematically and disproportionately denied entry to the work itself? Black and Brown people are more often than not culturized to embrace hard work as a means to success and that the harder we work, the more likely we are to achieve our goals. Sadly, our hard work does not negate the fact that we continue to face insurmountable barriers to achieve the same level of success as our White counterparts even though we are working just as hard or harder. When people of color are left bewildered as to why our hard work has failed us, the finger points to individual shortcomings and not the broken systems that continue to perpetuate racism. White Americans believe that their privilege by virtue of their skin color has nothing to do with their self-reliance and hard work that yields their success. If they can do it, so can everyone else. In reality, success can be achieved by all only when the rules equally apply to all.

Champions of inclusion understand the impact of color blindness and view seeing color as a gift that enables us to discover unique qualities, appreciate new perspectives, and better understand a person's lived experiences. Understanding the impact of color blindness means taking the blinders off and altering the way we think about race and racism. We must see color to develop the language needed to have courageous conversations about race and examine our biases. Inclusion requires developing new knowledge and new skills and exploring attitudes about race and racism while simultaneously applying what's learned.

Actions

Peel Back the Blinders to Color

Whether you were reared as a child to not see color or subscribed to color blindness as an adult to be perceived as egalitarian, nonracist, or becoming anti-racist, you've got some unlearning to do. Color-blind narratives can normalize inequity, disregard racial history, or attribute unfortunate or devasting life outcomes of melanated people to personal character flaws. These narratives must be challenged through the lens of inclusion. Commit to the work that increases understanding around racism. Racism isn't just about the intentional and individual acts of inflicting harm or discriminating against people based solely on skin color, but a system that individuals continue to perpetuate in any number of ways like remaining silent or tolerant when witnessing racist acts or minimizing the racialized trauma afflicting Black and Brown families when voicing "All Lives Matter" in place of "Black Lives Matter." Become part of the solution. Examine whether your narrative fosters curiosity about another's experiences, provokes self-reflection, or evokes personal accountability in maintaining the status quo. Learn to trust others' accounts of painful life experiences rather than minimize them to maintain comfort.

Understand Racism Through the Lens of the Oppressed

Color blindness is an unintentional act of racism designed to reduce perceptions of discrimination and increase perceptions of inclusion while deflecting historical oppression faced by people of color. It is not a stretch of the imagination that a White male of privilege would have a different point of view of what racism is than that of a Black or Brown man. The people who commit racist acts are not and should not be the ones who get to decide whether it was racist. Race is always at play during interactions to varying degrees no matter how sincere the attempt to not notice skin color. The racially oppressed don't have the luxury of not noticing skin color, especially when they are one of a few, the first, or the only. Racism is seen, felt, and heard and can come across unconsciously; thus, one must make the unconscious conscious and act with intention.

Seek Opportunities to Learn More

Our opinions and beliefs are shaped and colored by our experiences and what we feed our minds. Self-examination and education must be ongoing, as one-and-done is unrealistic. One book, one workshop, one blog, and one expert or self-proclaimed expert is not enough to learn the ugly truth of America's roots. Diverse sources are a requirement for diverse perspectives and the ability to move past defending racist behavior to a place of growth. There's no such thing as the perfect anecdote that will forever change one's perspective after a lifetime of racist ideals hidden behind color blindness. Apply what is learned. It will take courage to ask questions to better understand where and how personal behaviors contribute to the harsh realities of racism.

Action Accelerators

- **HBS.edu**: Harvard Business School, "Racial Color Blindness: Emergence, Practice, and Implications": www.hbs.edu/ris/ Publication%20Files/Racial%20Color%20Blindness_16f0f9c6-9a67- 4125-ae30-5eb1ae1eff59.pdf

- **UCLALawReview.org**: "The Indignities of Color Blindness," by Elise C. Boddie: www.uclalawreview.org/indignities-color- blindness

- **Forbes.com**: "The Case Against Racial Colorblindness in the Workplace": www.forbes.com/sites/hbsworkingknowledge/ 2013/01/20/the-case-against-racial-colorblindness- in-the-workplace/?sh=577f922036a6

Sources Cited

ClevelandClinic.org. "Color Blindness," my.clevelandclinic .org/health/diseases/11604-color-blindness#:~:text=The% 20gene%20responsible%20for%20color,because%20it%27s%20 a%20recessive%20trait

Nellie Tran and Susan E. Paterson. "'American' as a Proxy for 'Whiteness': Racial Color-Blindness in Everyday Life," *Women & Therapy*, 38:3-4, 341-355, DOI:10.1080/02703149.2015.1059216, 2015, www.tandfonline.com/doi/full/10.1080/02703149.2015.1059216

William Rogers. "Little Progress for Black CEOs in the U.S.," Statista.com, September 30, 2020, www.statista.com/chart/23060/growth-in-minority-executives

Create a Community of Allies

"The time when you need to do something is when no one else is willing to do it, when people are saying it can't be done."

—Mary Frances Berry

***

We've all endured the agony of defeat. We give something our best effort, and for one reason or another, our best did not yield the desired outcome. So, what did we do? We cut our losses and quit! For committed allies, quitting is not an option. The work is far too important. There are broken systems and policies that need to be dismantled and rebuilt, there are biases and prejudices to be addressed, there are lives to improve, and so much more that cannot happen unless we get involved. The work requires building allyship skills, competencies, and an awareness of the issues of the groups or individuals we advocate with and for. Our learning curves tend to be accompanied by unfortunate and unintentional outcomes that may cause us to second-guess our commitment. On our journey, we may discover that our well-intentioned speaking up on behalf of someone whom we believe was treated unfairly is met with resentment because we did not make space for them to speak up for themselves. Perhaps an affirming compliment shared with a person of color or of a different culture is shunned as they found it offensive. Things happen and we're not going to always get it right. There is no one-size-fits-all approach.

If you have started your allyship journey as recommended in Activity 13, "Aspire to Be an Ally," you already know that allyship is an ongoing, constant set of practices requiring endurance to be effective over the long term. By now you've probably discovered that being an active ally is an inherently uncomfortable thing to do. Knowing that you don't know what you don't know is scary and makes one vulnerable. We'll need courage and determination to stay the course. Allies need allies as a system of support to keep us motivated through mistakes, setbacks, and misunderstandings as we move toward more inclusive cultures. There will be times when we need a shoulder to cry on, a fresh perspective from another's point of view, or guidance on what to do next. While the work of allyship is rewarding, the stress and frustration of remaining diligent in standing up for inclusion, equity, and equality can become intense or even emotionally draining. The more allies we can partner with, the better our resilience and collective impact. Our allies have their own struggles, and we don't want to always call on the same ones for support while remembering that we are their allies when they need us. It's going to take a village, and everybody needs someone to lean on. When likeminded individuals come together for a common purpose, not only can we support one another but we can also accelerate change for the good of all. I've been told that the best way to eat an elephant is one bite at a time, and building a village of allies is much like eating the proverbial elephant. Start small by sharing your allyship journey with those closest to you and encouraging more people to join over time. Set a goal to add one community member a month. Reach out by communicating your why of becoming an ally, and why you think they would make a great one as well. Remember to include successes you've had thus far and how those made you feel and provide the lessons learned along the way to help them avoid the same mistakes. Create a cadence for connecting. Champions of inclusion choose not to be bystanders in the quest for inclusion and equality, and they collaborate beyond race and gender to create inclusive coalitions that can forge ahead despite the obstacles.

Actions

Develop a Recruitment Plan

You may have heard the saying if you fail to plan, you plan to fail. Having an outreach strategy, an attainable goal, and compelling success stories can encourage people to become change agents. Consider likeminded individuals in your organization from entry level to the executive level as well as those in your personal network. If applicable, engage employee resource group/business resource group (ERG/BRG) leaders and their members. As the community builds, incoming individuals can begin to recruit from their network.

Look for Gaps in Expertise

There will be times when a cause presents itself where allies are ill-equipped and unaware of the challenges, especially in situations relating to an individual who is the first, one of a few, or the only one like them. For example, gender-neutral restrooms are top of mind not only for organizations but for staff as well. There's a strong possibility that the existing community of allies may need guidance to work toward an equitable outcome and how best to support LGBTQ colleagues in the process. Don't shy away when you don't know what to do. Figure it out.

Ensure Allyship Community Is Diverse

This is not the time to be exclusive. Causes come in various shapes and levels of complexity. The most effective solutions come from varied perspectives and experiences. Work to ensure that the community of allies go beyond race and gender to include age, ability, religion, areas of expertise, culture, socio-economic background, and LGBTQ individuals.

Action Accelerators

▪ **TrainingMag.com**: "The 5 Types of Workplace Allies and How to Be One", by Angela Peacock: trainingmag.com/the-5-types-of-workplace-allies-and-how-to-be-one

- ■ **HBR.org**: "Be a Better Ally," by Tsedale M. Melaku, Angie Beeman, David G. Smith, and W. Brad Johnson: hbr.org/2020/11/be-a-better-ally

- ■ **YouTube.com**: *Want a more just world? Be an unlikely ally*, Dwinita Mosby Tyler: www.youtube.com/watch?v=Ruf6OdDRJSs

- ■ **YouTube.com**: *A simple way to be an LGBTQ+ ally*, Keegan Thoranin, TEDxLFHS: www.youtube.com/watch?v=eCXQiAG_6cg

Pick a Ship and Get on It

"Feelings of worth can flourish only in an atmosphere where individual differences are appreciated, mistakes are tolerated, communication is open, and rules are flexible—the kind of atmosphere that is found in a nurturing family."

—Virginia Satir

***

The more I learn about the impact of "ships" in the creation of more equitable and inclusive workplaces and society as a whole, the more excited I become by the magnitude of change we can produce. When we roll up our sleeves and embark upon a journey of mentorship, sponsorship, relationship, or allyship, we can improve lives and change systems. Look at the advancements we've made in societal dysfunctions ultimately affecting the workplace:

- The Marriage Equality Act resulted in more inclusive benefits for the LGBTQ community.
- Pay equity legislation prompted more employers to implement fair pay policies.

- The American with Disabilities Act requires employers with 15 or more employees to provide reasonable accommodations for individuals with disabilities creating equal opportunity for employment.

- Affirmative Action aims toward anti-discrimination and equal opportunities for historically excluded groups when making college admission and employment decisions.

We've come a long way, and we have a very long way to go. Ships create opportunity to connect with causes and individuals we care about and can empower us to be the change we want and need to see either on the front lines or behind the scenes. Anyone at any age regardless of race, gender, or ability can make an impact in a ship. I've heard heroic stories of grade-school children who got on the ally-ship when they crafted masks for the homeless during the height of covid-19 or shut down a bully for harassing a classmate for wearing eyeglasses. I first jumped on in the second grade when I secretly grabbed an extra pair of my mittens and a scarf to gift a schoolmate on the playground who shivered every recess on chilly days. Even children can relate to circumstances that don't necessarily affect or involve them but choose to step up anyway, sometimes in the face of grave danger. Consider the story of Steven Stayner who was kidnapped as a seven-year-old, held captive, and sexually abused by his abductor for seven years before escaping and freeing a younger kidnapped boy.

When we witness suffering, injustice, and inequities or are simply aware and have a desire to be a part of the solution, there is a ship waiting for us to get in. All ships are not the same. Each presents a unique set of circumstances and varying amounts of effort, energy, time, and sometimes financial resources. Here are more details:

- **Allyship:** As explained by Sheree Atcheson, "The direct action of an ally and the specific work done by an ally to learn (or unlearn) the experiences of different communities whilst using their privilege to shift the dial for these groups."

- **Sponsorship:** As explained by Michelle Silverthorne, "Sponsorship is centered on advancing marginalized groups in the workplace to positions of power. The sponsor, therefore, has to be someone with significant influence in the organization, and who will advocate for, protect, and fight for the career advancement of the employee that they sponsor."

Figure 42.1: Allyship

Figure 42.2: Sponsor and protege

- **Mentorship:** As explained by Berkley Labs Computer Sciences, "Mentorship is a protected relationship in which a more knowledgeable or experienced person guides and nurtures the professional development or growth of another, outside the normal manager/subordinate line management. Its focus goes beyond learning specific competencies or tasks and builds a climate of trust so the mentee can feel secure to seek advice on issues impacting their professional success. Mentorship achieves its goals primarily by listening with empathy, sharing experiences, developing insight through reflection, and encouraging the mentee to take action towards the achievement of self-driven goals. As such the relationship is mentee-driven, with a clear definition of goals and expectations that are mutually agreed with the mentor. It is attentive to the mentee's values and needs and respectful of the mentor's time, resources, and experience."

Figure 42.3: Mentor and mentee

- **Workplace Relationship:** As defined by Dictionary.net, "Workplace relationships are unique interpersonal relationships with important implications for the individuals in those relationships, and the organizations in which the relationships exist and develop. Workplace relationships directly affect a worker's ability and drive to succeed. These connections are multifaceted, can exist in and out of the organization, and be both positive and negative. One such detriment lies in the nonexistence of workplace relationships, which can lead to feelings of loneliness. Workplace relationships are not limited to friendships, but also include superior-subordinate, romantic, and family relationships."

Figure 42.4: Relationship at work

All ships will require patience, empathy, trust, and respect. Others, like sponsorship, require an individual to be in an executive-level role with power and influence, while mentorship requires a level of expertise in specialized areas to be beneficial to the mentee. Common causes that ships support include but are not limited to the following:

- Career advancement
- Professional development, skills development
- Racial, gender, people with disabilities, and LGBTQ equity
- Psychological safety
- Personal growth
- Imposter syndrome
- Advancing diversity

There are a few other ships to make note of on your inclusion journey. As marketplace demands change and workforce strategies adjust to attract and nurture new talent from diverse backgrounds, organizations may leverage internships, apprenticeships, and returnships.

- **Apprenticeship:** As explained by TheBalancedCareer.com, "An apprenticeship is a career path where individuals can obtain paid work experience, classroom instruction, and an industry and nationally recognized credential. For an employer, an apprenticeship program is a means to train, develop, and prepare their workforce. An apprenticeship program is comprised of five components: paid employment, work-based learning, classroom learning, mentorship, and professional credentials."

■ **Internship:** As explained by Careers.UMBC.edu, "An internship is a professional learning experience that offers meaningful, practical work related to a student's field of study or career interest. An internship gives a student the opportunity for career exploration and development, and to learn new skills. It offers the employer the opportunity to bring new ideas and energy into the workplace, develop talent and potentially build a pipeline for future full-time employees. A quality internship consists of a part-time or full-time work schedule that includes no more than 25 percent clerical or administrative duties, provides a clear job/project description for the work experience, orients the student to the organization, its culture and proposed work assignment(s), helps the student develop and achieve learning goals, and offers regular feedback to the student intern."

■ **Returnship:** As explained by TheBalanceCareer.com, "Returnships are return-to-work programs for professionals who have left the workforce, typically to care for young children or family members. These opportunities are paid internships targeted at workers with significant working-time gaps in their resumes. Returners may be working parents, caregivers, or other professionals who have had to press pause on their careers for a period of time. These programs may also provide an opportunity to change careers. Workers who previously held positions in fields that are in decline may use returnships to transition to fields with more solid occupational outlooks. Some programs include 'underemployed' as part of the eligibility criteria. These programs can also provide an on ramp to private-sector employment for veterans who are transitioning out of their military careers."

Aligning ourselves with the success of interns and apprentices via allyship, mentorship, or relationship will help them feel more included as many will be unprepared for the struggles they'll face for being different. Champions of inclusion consistently inhabit ships to fulfill their commitment to changing the status quo.

Actions

Don't Limit Yourself to One Ship

With so many options, there's always a ship waiting on us. With each requiring varying levels of effort and energy, some of us may have bandwidth for more than one. Co-workers who are one of a few, the first, or the only one with their identity create opportunities for ships to happen. Refer to Activity 33, "Support the Firsts, Fews, and Onlys." Look around the organization for opportunities to be deployed for a cause or to join forces with others. The allyship journey may include mentoring or sponsoring a deserving individual, so it's important to maintain balance. Quality over quantity.

Do Your Homework

Ships usually require working cross-culturally, understanding the issues related to the cause, and knowing what we bring to the table. As we endeavor toward quality ships, prepping in advance will increase the chances of meaningful connection to the individual or cause and set us up for success. Leverage the skills we've worked on so far: inclusive language, correct pronoun usage, and cultural competence. Review Activity 5, "Experience Other Cultures;" Activity 12, "Strive to See the Whole Person;" Activity 19, "Invest in the Success of Others;" and Activity 20, "Lead Change One Word at a Time," for guidance.

Understand Personal Preferences

Mutually beneficial outcomes are achieved when we work in alignment with natural strengths and areas we're passionate about. It's important to be aware of where we can make the most impact and our competency in which to do so. Now that you are more aware of the various ships, evaluate where you would like to start or which ship to embark upon next.

Action Accelerators

- **Blog.Hubspot.com**: "How to Be an Amazing Mentor in 10 Ways, according to HubSpot Managers," by Martina Bretous: `blog .hubspot.com/marketing/mentor-tips-positive-impact`

- **StrategyandBusiness.com:** "How to Be a Great Sponsor," by Wanda Wallace: www.strategy-business.com/blog/How-to-be-a-great-sponsor

- **YouTube.com:** *Your words have the power to end suffering of LGBTQ youth*, Kat Clark, TEDxTauranga: www.youtube.com/watch?v=2zuqUpiM26c

- **CultureAmp.com:** "Four Practices for Being an Equality Ally": www.cultureamp.com/blog/four-practices-for-being-an-equality-ally

- **Indeed.com:** "How To Build Good Working Relationships at Work, by Indeed Editorial Team": www.indeed.com/career-advice/starting-new-job/how-to-build-good-working-relationships

Sources Cited

"What is Mentorship?" cs.lbl.gov/diversity-equity-and-inclusion/csa-mentoring-program/what-is-mentorship

Sheree Atcheson. *Demanding More*, page 139

Michelle Silverthorn. *How to Change the Workplace for Good*, Authentic Diversity, page 137

Alison Doyle. "Best Apprenticeship Programs," TheBalanceCareers.com, October 1, 2020, www.thebalancecareers.com/best-apprenticeship-programs-5080350

"What is an internship?" Careers.UMBC.edu, careers.umbc.edu/employers/internships/what-is-an-internship

43

Say Their Name

"A person's name is to him or her the sweetest and most important sound in any language."

—Dale Carnegie

***

What's in a name? For many, our name is representative of our heritage, culture, or something of significance to our parents. I once worked with a man named Paris and believed his name was more feminine than masculine. I was surprised when he agreed with me. He beamed when he told me that that's where he was conceived while his parents vacationed there. In his mind, that vacation was more special than any other they would take. Feminine or not, he was proud of his name and why he had it. Paris was lucky. He had a name that is easily pronounced. My first name is unusual and is often mispronounced. I have met only a handful of individuals who share my name. The phonetic enunciation is "ee-VET," and it is of French origin. I grew to like my name over the years because it was different, and I got a kick out of informing people that it was French. I used to marvel at being Black and having a French name.

As a kid, somehow that made me feel cool. People who get it right on the first attempt tend to pose it as a question, "Yvette?" rather than saying it with confidence. For some reason, people find it necessary to sound out the letter Y. Classmates struggled far less saying my name than faculty and required correcting only a few times before nailing it every time. My high school history teacher butchered my name as "Ya-Vetter" every single day no matter how many times I corrected her. It's exhausting to fight a battle that you'll never win, so I stopped correcting her and decided to try something else. My grandmother used to say, "It's not what they call you—it's what you answer to." Following her sage wisdom, I decided not to respond when that teacher mispronounced my name. As a 16-year-old, not answering to an authority figure is a big mistake. I stood my ground for nearly a week until it almost got me suspended. I guess she got fed up with my rebellion. She won the battle and pronounced my name incorrectly for the entire school year. The Catholic high school I attended was diverse, but the faculty was 100 percent White. The faculty members who habitually mispronounced my name made me feel like an outsider. I wondered why they didn't care enough to get it right. The irony of educators unwilling to learn how to correctly pronounce students' names did not make sense. By contrast, students were forbidden to mispronounce the teacher's name. It was considered a sign of disrespect to say Mr. or Ms. (fill in a last name) and not get it right. When substitute teachers were called in, they genuinely struggled with non-Anglo-American sounding names and, sadly, never once asked the appropriate student whether they got it right or for assistance in pronouncing it correctly. But when your name is mispronounced most of your life, you get used to hearing it annihilated in more ways than you thought possible and take responsibility to ease the discomfort of the offender. Imagine the roll call going in a smooth rhythm of saying a name, pausing for a second to look for or hear a signal: Crystal, Josh, Steven, Mary, Brittany, and then the next sound is "uh" or "hmmm" or my favorite "I'm not going to say your name; I'll just spell it." In school, you knew when it was your name that broke the rhythm as you consistently heard the name preceding yours, so jumping in with a quick response restored the rhythm and reduced the awkwardness. A post on FastCompany.com paints a clear picture of the lived experience and the impact of enduring a lifetime of people refusing to learn and correctly pronounce someone's name:

Many people don't know this, but my full name is Madhumita Mallick. I have spent much of my life trying to hide it. While my name was a source of pride, part of my identity, and represented my heritage, it was also a source of anxiety, embarrassment, and shame.

In the third grade, my teacher couldn't pronounce Mita, never mind the full name. She thought it would be fun to call me Pita ("Like pita pockets," she said). In college, my closest friends gave me the nickname Mahu after our calculus professor called me Mahu Mallick. He didn't want to be corrected.

"Honestly, go by Mita Mallick," the career counselor coached me, striking through my name in red on my résumé. "No one can pronounce this, you won't get callbacks." But it was my full name. I wanted to reclaim and embrace my name, which was what my expensive liberal arts education taught me.

At my first corporate job, I found myself back in those first days of school. My full name was simply too confusing for people.

"I thought your name is Mita? I can't find you in the distribution list." "I walked around looking for you and didn't see Mita on any of the cubes." "Why don't you just go by Mita?"

Then there was my manager who thought he had the best suggestion—"Mohammed." From that moment on, he called me by this left-field title, simply because he could.

"Mohammed, did you pull that Nielsen data the team asked for?" "Mohammed, can you join us for the 4 p.m. call?" "Mohammed, make sure the agency knows to dial in for the kickoff."

I responded to a name that wasn't mine for close to six months before I left. And before I did, I wish I had just said one thing, "Call me by *my* name."

One of the biggest microaggressions that can take place is the repeated mispronunciation of someone's name. Or in my case, completely changing someone's name. When my boss created this new nickname for me, that served as a form of bullying and harassment.

Not much has changed over the decades. Those of us with unusual and unfamiliar names still find that the onus is on us to ease the discomfort of those refusing to learn to say our names correctly. It's understandable when we first encounter a name to initially mispronounce it. It's the unwillingness to learn the pronunciation that's the issue. I find that I am less offended when people who mispronounce my name are making a sincere effort to learn how to say it correctly and more offended when they continuously mispronounce it after they have been corrected several times. It's like saying "I don't care how you say your name; I'm going to call you what I want to call you." BBC.com reports:

Xian Zhao, a post-doctoral fellow at the University of Toronto whose research focuses on ethnic name pronunciation, says that although many people don't realise it, habitually pronouncing an unfamiliar name incorrectly is a form of implicit discrimination. It sends a message that "you are minimal," says Zhao. "You are not important in this environment, so why should I take time and my effort to learn it?"

If you are a name butcher, it is incumbent upon you to learn how to correctly pronounce the names of others and not force them to surrender part of themselves for the sake of your comfort. The practice of making the attempt, botching it, and then glossing over it as if nothing happened has harmful effects. Consistently mispronouncing a co-worker's name has a big impact over time, and so does saying it correctly. For individuals who experience discrimination based on their identity, persistent mispronunciation of their name can feel like erasure or insulting. People invest the time to learn new languages. Why not names? Learning to correctly pronounce names requires far less effort and energy than learning a new language, and the return on the investment yields more inclusive cultures and a sense of belonging. The person who has mastered the correct pronunciation is perceived as someone who respects the feelings and identities of others while the recipient feels more valued and seen. Everybody loves it when their name is spoken correctly. It is a big part of who we are and expressions of our individuality. We may have co-workers who have chosen a preferred name that better aligns with who they truly are, and they abandon the use of their government name altogether in the workplace and social settings or may, in fact, legally change their name. As the workforce becomes more diverse, so will the names of individuals we work with. Learning to say unfamiliar names or remembering when someone has a preferred name will require practice and patience with intention. Champions of inclusion build and enhance relationships by approaching individuals' names with the intent to create a sense of belonging. We are intentional about not only working toward saying people's names correctly but also spelling it in the way that they spell it in written communications and using preferred names. While Brian is commonly spelled with an *i*, others may spell it as Bryan. Does your colleague spell her name Teresa or Theresa, Michelle or Michele? It's important to get it right. The three basic rules for being respectful of people's names are the following:

- Pronouncing it correctly
- Spelling it as the name owner spells it
- Remembering to use preferred names

Normalizing our approach to getting names right can diminish the discomfort and embarrassment over not knowing how to say something. Inclusion requires a learning mindset, and we must learn how people want to be addressed and respond accordingly.

Actions

Read the full article cited by `FastCompany.com`. It advises:

ASK PEOPLE FROM THE START HOW TO PRONOUNCE THEIR NAME

Even if it's the second or third time you are meeting them, you can say, "It's important for me to pronounce your name correctly, and I know I asked you last time as well. But can you spell your name and say it for me phonetically please?" This not only shows that you want to be able to pronounce someone's name correctly, but also signals to them that you care deeply and want to build a mutually respectful relationship.

ASK OTHERS HOW TO PRONOUNCE YOUR COLLEAGUE'S NAME

Sometimes, we may feel uncomfortable or embarrassed to ask the person directly or too much time has passed. In these cases, don't feel sheepish to ask another colleague or a friend how to pronounce the person's name correctly.

CORRECT OTHERS WHEN YOU HEAR SOMEONE'S NAME MISPRONOUNCED

If someone has mispronounced a colleague or a friend's name, please be an advocate and stand up for them. "I am not sure if you realize this. I have heard you on a few occasions mispronounce Mita's name. The correct pronunciation is 'mee-ta.'"

USE ONLINE TOOLS TO CORRECTLY PRONOUNCE SOMEONE'S NAME

One of the best innovations this year is LinkedIn's name pronunciation feature. By using the tool, you can listen to the recording on the person's profile to hear how they say their name. So, you don't mispronounce or try to figure out how you avoid saying their name in a conversation completely.

DON'T ASSIGN NICKNAMES WITHOUT THEIR PERMISSION

If someone is called Jennifer, don't call them Jen without asking. Don't assume Matthew is Matt, especially if the individual hasn't given you permission to refer to them by a nickname. And please don't create a separate, unrelated nickname for them (in my case, "Mohammed"). Moreover, if you see someone creating an unwanted nickname for someone else, please intervene. Prevent the nickname from being used further and step up as an advocate. When you see something, say something.

Create a Cheat Sheet

Practice makes perfect. When someone shares how to say their name, write down the phonetic spelling and practice it until it's easily spoken correctly.

Double-Check Before Sending

Spelling a name right is just as important as saying a name right. Don't click Send until you are certain that the name is spelled as the owner spells it. Christine may spell her name "Cristine."

Prioritize the Preferred Name

When individuals have a preferred name, it's a sign of respect to call them by it, and it fosters a sense of belonging. When someone shares a preferred or chosen name, make using that one a priority in communications and interactions.

Action Accelerators

- **NPR.org**: "What Listeners Told Us About the Importance Of Getting Names Right," Clare Lombardo: www.npr.org/2021/05/02/989609197/what-listeners-told-us-about-the-importance-of-getting-names-right

- **HBR.org**: "If You Don't Know How to Say Someone's Name, Just Ask," by Ruchika Tulshyan: hbr.org/2020/01/if-you-dont-know-how-to-say-someones-name-just-ask

- **YouTube.com**: *Getting it right; why pronouncing names correctly matters*, Gerardo Ochoa, TEDxMcMinnville: www.youtube.com/watch?v=58tDCaEWfHI

- **ADP.com**: "10 Best Practices for Using Preferred or Chosen Names at Work," by Brett Daniel: www.adp.com/spark/articles/2022/06/10-best-practices-for-using-preferred-or-chosen-names-at-work.aspx

Sources Cited

FastCompany.com. www.fastcompany.com/90533252/why-this-workplace-microaggression-does-more-harm-than-you-realize

Zulekha Nathoo. Why Getting a Name Right Matters, BBC.com, January 11, 2021, www.bbc.com/worklife/article/20210108-the-signals-we-send-when-we-get-names-wrong

44

Master Small Talk Across Cultures

"Cultural integration doesn't happen by you boasting about your culture, it happens by you coming forward enthusiastically to learn about another culture."

—Abhijit Naskar, *Mücadele Muhabbet*

***

Where there is relationship and human connection, there is opportunity for inclusion to thrive. More often than not, relationships begin with small talk—those seemingly meaningless conversations about the weather or last night's game helps us to connect—and authentic connection improves interactions. Everybody engages in small talk. Some of us have mastered it, while others despise it with a passion and prefer that one just get to the point—please! For some it's a common courtesy and helps to avoid awkward and uncomfortable silences at the elevator, watercooler, stock room, or lunch table. Yet, others don't mind the silence as they think small talk is considered a waste of time or gets in the way of meaningful conversation, or they just simply are not good at it and choose to avoid it at all costs. I started my career in an era where small talk was frowned upon by management. If caught engaging in small talk, I was always reprimanded to get back to work. Small talk was the equivalent of wasting time. In some organizations, that belief persists. Whether we realize it or not, the ability to engage in small talk is an essential social skill that can make us feel more secure and connected, especially when we're working remotely. A post by Berrett-Koehler Publishers shares several reasons that small talk is important:

- It helps to establish trust.
- It lays the groundwork for specific requests.
- It may yield important information.

Most sources define small talk simply as chitchat, or short conversations on meaningless topics. My preferred definition comes from UniversalClass.com. They define it as follows:

Small talk is defined as the use of casual conversation about relatable topics with the goal of getting a sense for another person and beginning to establish rapport. Small talk will vary across different contexts. Small talk is often used around people that you do not know very well. In many ways, small talk is a defense mechanism to cope with the universal anxiety people experience when conversing with those they do not know well. Alternatively, small talk can also become more than just a defense mechanism, it can become a highly useful skill in the professional world. Small talk can help in meeting other people, building working rapport, and developing wide networks.

Typically, Americans are more inclined to engage in small talk as a means to establish rapport. Often, without realizing it, unexpected encounters and small talk conversations with co-workers spark ideas, foster collaboration, or create opportunities to vent frustrations—all of which can lead to improved performance. However, small talk is not a universal means in which to establish rapport and strengthen connection as its norms vary across cultures. No two cultures are exactly alike. An article from HBR.org says:

In many places around the world, it is unbecoming to engage in trivial banter about the weather or the commute to the office, or to glide from one topic to the other in a lighthearted fashion. In China, for example, people can be quite guarded and protective with personal information among people they do not know well—especially people they perceive to be in competition with

> for limited resources. The logic is that if people reveal personal
> information, it could be used against them in some way and lead
> to a strategic disadvantage.

With many organizations and institutions steadily increasing diverse populations, chance encounters are inevitable with people who are different from us, and we may find ourselves to be "fish out of water" when attempting to connect across cultures. Avoid jumping to the conclusion that they are rude, aloof, uninterested, or that something is wrong with *them*. It's human nature to be wary of striking a conversation with a stranger or someone whom we barely know. While small talk may not be universal, a sense of belonging and relationships are. Connect by readjusting expectations, learning cultural norms, and taking a longer-term view to rapport building. Expand small talk skills to take into consideration cultural attitudes toward appropriate topics and conversational tones. For Americans, asking "How are you?" can be either a mere greeting to which you respond, "Fine and you?" and move on or choose to respond with an explanation of your current state of being. Take into account that in some cultures asking "How are you" is an invitation for an engaging discussion or considered a personal question that invades their privacy. There are cultures that don't engage in small talk at all until a trusting relationship has developed, while others deem it unbecoming to engage in trivial chitchat. For Americans, small talk about politics and religion in the workplace is taboo, while other cultures enjoy a healthy debate on the topics. Inclusion demands that we not expect diverse individuals to assimilate to the status quo but make space for people to engage where we're each comfortable. Champions of inclusion continuously improve upon cultural competency skills (review Activities 3, "Create New Habits," 12, "Strive to See the Whole Person," and 30, "Make the Workplace Safe for Everyone to Be Themselves") that result in meaningful small talk as a foundation to deeper connection. Small talk may be small, but don't discount it based on the name.

Actions

Get Curious

Consider colleagues from other cultures and previous interactions with them. Expand on those by arming yourself with knowledge on their cultural norms. Review tourism sites and brochures as they tend to provide tremendous detail on how the locals live and make the most of enjoying the culture while visiting there. Observe their workspace setup. Take note of plants, photos, and other décor that may give insight needed for meaningful small talk.

Be Agile in Your Approach

There are cultural norms for domestic as well as international colleagues. If you are part of a global team or consistently interact with internationals, you may find that, for instance, Japanese Americans may be easier to engage in small talk than those co-workers who reside in Japan and are not Japanese Americans. The strategies needed to connect with colleagues from different cultures in your own country vary from those when connecting with colleagues across continents and time zones.

Create Space to Connect on a Personal Level

When we don't know others well, we tend to fill in the blanks with assumptions that are predicated by bias and stereotypical beliefs. It's just human nature. Rather than jump to conclusions when small talk is met with resistance, take a step back and become vulnerable. Let folks know that you're not the best at getting to know others but are interested in getting to know more co-workers better. Ask them if they would be okay to share how they usually get to know people. Asking if they would be okay with it leaves space for them to say no—though that's highly unlikely. But who knows? If nothing else, we will either learn from them how best to connect or minimally continue to acknowledge their presence with a warm greeting at each encounter and move on. Continuous warm greetings will build trust over time and can open the door to small talk.

Action Accelerators

- **Nasdaq.com**: "The Understated Importance of Office Small Talk," by Chris Morris: www.nasdaq.com/articles/the-understated-importance-of-office-small-talk-2021-02-26
- **Phys.org**: "Office small talk has a big impact on employees' well-being, study finds," by University Exeter: phys.org/news/2020-06-office-small-big-impact-employees.html

- YouTube.com: *Cross cultural communication,* Pellegrino Riccardi, TEDxBergen: www.youtube.com/watch?v=YMyofREc5Jk&t=780s
- Northeastern.edu: "How to Improve Cross-Cultural Communication in the Workplace," by Tim Stobierski: www.northeastern.edu/graduate/blog/cross-cultural-communication

Sources Cited

UniversalClass.com. "Mastering Conversation: The Art of Small Talk,"www.universalclass.com/articles/self-help/mastering-conversation-the-art-of-small-talk.htm

Jeevan Sivasubramaniam. "5 Ways Small Talk Serves a Big Purpose," Ideas.BKconnection.com,ideas.bkconnection.com/5-ways-small-talk-serves-a-big-purpose

Andy Molinsky and Melissa Hahn. "Building Relationships in Cultures That Don't Do Small Talk," HBR.org, April 8, 2015, hbr.org/2015/04/building-relationships-in-cultures-that-dont-do-small-talk

Enable Uncovering

"To be yourself in a world that is constantly trying to make you something else is the greatest accomplishment."

—Ralph Waldo Emerson

***

The ability to live life in the truest version of ourselves is empowering. All too often we allow what others think about and expect of us to get in the way of showing all aspects of who we are. When we find ourselves in environments that don't welcome key facets of who we are, we are compelled to cover them to feel safe. The term *covering* has gained popularity over the years from the body of work done by Kenji Yoshino, legal scholar and author of *Covering: The Hidden Assault on Our Civil Rights*. He defines it as "editing, modifying, and downplaying a known stigmatized identity so one can operate effectively at work. The rationale being that while you can't change yourself you can change how yourself shows up to work." For some, covering may mean leaving the best part of themselves at the workplace door. When that happens, employers and teams are unable to harness all the talents we have to offer, and we miss out on the opportunity to maximize our potential. Enabling the covered to uncover requires that we create psychological safety and use it in our day-to-day interactions. This is also vital to the development of inclusive cultures. We feel psychologically safe when we can be our authentic selves without fear of repercussion just for being who we are. Safety is one of those things that we may take for granted in the workplace, especially when the organization's branding espouses its commitment to diversity, equity, and inclusion. We would like to believe that the environment values difference and is free of harassment and discrimination for all individuals regardless of age, gender, race, ability, religion, and LGBTQ identity so that we don't have to cover parts of ourselves to be accepted. While many companies to varying degrees have policies and practices prohibiting discrimination, the truth of the matter is that many individuals don't feel safe revealing aspects of themselves for fear of discrimination and being stigmatized. Representing known characteristics that may be frowned upon can jeopardize careers, ruin opportunities to advance, or worse make us a target for being bullied or harassed. According to the Deloitte survey "Inclusion survey: Uncovering talent," "sixty-one percent of participating employees cover their identities in some way at work." Imagine hiding, downplaying parts of yourself, or pretending to be someone that you're not for the entire time that you're at work or participating in work-related activities. Even the simplest of interactions may raise stress and anxiety levels as one focuses more on *guarding* themselves than *being* themselves. Dr. Helen Ofosu, MA and PhD in applied social psychology, shares in a post on IOadvisory.com that:

In addition to covering at work taking a toll on an employee's productivity and creativity, it may also take a toll on their mental health. Although it's common to act more reserved and professional at work, when the gap between your personal self and your professional/public self is large, it can be draining. When you feel like you're constantly acting, it can be tiring and difficult to sustain over the long-term.

People's perceptions of what defines the norm is very real. Our attitudes and beliefs show themselves during interactions as sexist, racist, or anti-LGBTQ, and other microaggressions persist. Negative bias and stigmas will cause White people to hide that they are married to a person of color, individuals with disabilities who function better in wheelchairs to opt for a cane to make the disability less noticeable, or practicing Muslims to pray in secret rather than risk being noticed in common areas.

The Deloitte survey defines covering in four dimensions:

Appearance based—covering concerns how individuals alter their self-presentation—including grooming, attire, and mannerisms—to blend into the mainstream. For instance, a black woman might straighten her hair to deemphasize her race.

Affiliation based—covering concerns how individuals avoid behaviors widely associated with their identity, often to negate stereotypes about that identity. A woman might avoid talking about being a mother because she does not want her colleagues to think that she is less committed to work.

Advocacy based—covering concerns how much individuals "stick up for" their group. A veteran might refrain from challenging a joke about the military, lest she be seen as overly strident.

Association based—covering concerns how individuals avoid contact with other group members. A gay person might refrain from bringing his same sex partner to a work function so as not to be seen as "too gay."

Workplace culture becomes conducive to uncovering when we do things differently, do different things, or stop doing things altogether. Doing things differently may mean interrupting biased thinking and stereotypical beliefs rather than allowing them to play out during interactions and decision-making. Doing different things may mean joining an affinity group (i.e., individuals with disabilities) as an ally to better understand the challenges and become more aware of how personal actions and attitudes diminish someone's experience in the workplace. Stop doing things altogether may mean no longer telling jokes that cause or perpetuate unfavorable perceptions of marginalized groups. Co-workers should not be pressured or made to feel pressured by other co-workers to modify or downplay aspects of their identity so that they can feel more comfortable around them. My grandmother always seemed to know when I wasn't being myself and, in those times, she warned, "What goes on in the dark will always come to light. You may not know when or how; it just always does." What that said to me is that in reality, I wasn't covering anything at all. She knew the truth—the real me—the one who I refused to reveal (see Figure 45.1). Parts of me were living in the shadows, and I was embarrassed by them. I wanted to live up to her expectations and not disappoint her with the truth. What I didn't realize at the time was that the warning came from a place of love. She knew the devasting effects of being uncovered on someone else's terms and desired better for me. She was right about the light. When the light of truth showed up, I was usually blindsided and sometimes with devastating consequences as I thought I was doing a darn good job in keeping things under wraps. Everyone deserves to live their truth when it's not harmful to others. Diminishing motherhood should not be a requirement for career progression. Nonbinary individuals should not be pressured to choose a feminine or masculine persona because their ambiguity makes others uncomfortable. No one truth should be held superior to another. Champions of inclusion are aware of the consequences of covering and the benefits of uncovering and let people know that fully expressing their differences are welcomed in their presence.

Figure 45.1: The me I want you to see

Actions

Develop Trust

We are inclined to reveal more of ourselves to individuals we trust. When we demonstrate our trustworthiness, trusting relationships emerge. Review Activity 25, "Foster an Environment of Trust," for guidance.

Check Yourself

Biased and stereotypical beliefs are primary culprits to covering. Examine your beliefs around age, ability, race, gender, religion, and the LGBTQ community. Work to eliminate, mitigate, and interrupt biased thinking with an inclusive lens. If you have not done so, take the IAT—implicit association test—to help identify biases. Review Activity 12, "Strive to See the Whole Person."

Let Vulnerability Show

Most folks don't believe themselves to be racist, sexist, homophobic, ageist, ableist, or a religion antagonist, but words and actions can cause one to be perceived as such. We may find out when we've been called out for committing a microaggression, or we can choose to let our vulnerability show when we proactively ask for feedback. Create a safe space to have conversations with individuals about what they have experienced or observed that gets in the way of them being able to be themselves in your presence. Collaborate on ways that you may shift or improve perceptions. Review Activity 28, "Reframe Difficult Conversations on Polarizing Topics," for guidance.

Action Accelerators

- **Podcast**: *Your Brain at Work*, "Inclusion, Covering and Authenticity: Interview with Global Expert Kenji Yoshino": www.stitcher.com/show/your-brain-at-work/episode/inclusion-covering-and-authenticity-interview-with-global-expert-kenji-yoshino-81947224

- **Forbes.com**: "The Difference Between Diversity and Inclusion and Why It Is Important to Your Success," by William Arruda: www.forbes.com/sites/williamarruda/2016/11/22/the-difference-between-diversity-and-inclusion-and-why-it-is-important-to-your-success/?sh=3687eeb25f8f

- **Catalyst.org**: Infographic. "What Is Covering?": www.catalyst.org/research/infographic-what-is-covering

- **QuantumWorkplace.com**: "9 Strategies to Create Psychological Safety at Work," by Shana Bosler: www.quantumworkplace.com/future-of-work/create-psychological-safety-in-the-workplace

Sources Cited

Deloitte.com. "Inclusion survey: Uncovering talent," www2.deloitte
.com/us/en/pages/about-deloitte/articles/covering-in-
the-workplace.html

Dr. Helen Ofosu. "Covering at Work: The Pros and Cons of Being
Ourselves at Work," IOadvisory.com, October 27, 2018, ioad
visory.com/being-ourselves-vs-covering-at-work

INTERVIEW: TELLING IT LIKE IT IS. . .TO GET WHERE WE NEED TO GO

Covering for Gain Pales in Comparison to What Is Lost

Jeremy (not his real name), a light-skinned, straight-haired, racially mixed Black man, born to a Black mother and White father, is light enough to pass for White. While in the sixth grade, his parents divorced, and he eventually lost contact with his father. His mom raised him in a racially integrated suburb of Chicago, and White people assumed that he was one of them until they discovered that the Black woman who accompanied him on every outing was not his nanny but, in fact, his mother. The distinct contrast in their appearance was never a concern for him, but others made it their concern when they expressed utter surprise at the discovery or made it a point to emphasize how they looked nothing alike. The polite White people asked whether he was adopted; others jested "C'mon, you're not his real mother." As a young kid, it was terrifying to think that the woman you've loved and called "mom" all of your life may not be your real mom. The slights continued throughout high school and negatively impacted his self-esteem. His mother's love kept him grounded. He was a proud son and proud of his Black heritage. It was in high school that he became more aware of the challenges of being Black. The experiences of his White classmates versus his Black ones were starkly different. His Black classmates were unfairly and disproportionately singled out for class disruptions and received more detentions when late for class. He noticed how they were followed around stores while he roamed freely. As he matured, he became painfully aware of the risks associated with being a person of color in America, especially an African American male. The fear of not surviving a mere traffic stop with police, stereotyped as violent, higher hurdles to career success, or simply enjoying a walk in the park caused Jeremy to daydream about what life would be like if he shed his Blackness to pass for White. By senior year, he started to distance himself from his Black friends in hopes of avoiding the perils of a Black man. At 19, he was accepted to a predominantly White university in Boston on an academic scholarship. He arrived on campus identifying as a White man. In his racial ambiguity, he floated among the student body as White and was intentional to not associate with other Black students outside class or otherwise make meaningful connections that had always come so easy. Covering his Blackness and rejecting the Black community, he disappeared into a White world. He altered his speech and mannerisms to appear more White—choosing to impersonate what he remembered about his dad. He vanished from social media and changed his style of dress. He could not muster the courage to share his decision with his mother and worked hard to keep her identity a secret. Recurring nightmares that she may pay him a surprise visit often led to panic attacks. The shift was difficult and painful, but he was determined to muscle through it. He put on Oscar award-winning performances pretending to know little of Black culture and hiding his longing to join Black classmates when they went out for soul food and partied to hip-hop music. When hanging with White students, they were often very open with their racist views and defended the actions of police and campus security in their treatment of Black students. The inability to advocate and educate them on the issues crushed him emotionally. He could feel his mother's disappointment or even disgust although she was not in the room. He recalls the return to campus after his first Spring Break. He watched everyone be themselves. They had stories, pictures, videos, and laughs over social media posts. He felt like an outsider looking in as he had only derivatives of the truth to share. In his isolation, he missed his mom. The life and love she had given him would soon be a distant memory. Deep down he felt that some Black people knew he wasn't White and dreaded that someday he would be called out. Next to not surviving an encounter with the police, being called out was his biggest fear. He was convinced that living safely and successfully in America would be impossible without the privileges that come with being White. As graduation loomed closer, he finally decided to break the news to his mother. He couldn't risk her attendance at the ceremony. He graduated with honors, accepted a job in Silicon Valley, and lives his life as a White man. He is no longer covering but passing. Years later, he continues to reflect on whether the decision was worth it. He would tell you that "while the gains were great, the losses were greater."

46

Diminish the Effects of Microaggressions

"It has been said, 'time heals all wounds.' I do not agree. The wounds remain. In time, the mind, protecting its sanity, covers them with scar tissue and the pain lessens. But it is never gone."

—**Rose Fitzgerald Kennedy**

***

Microaggressions hurt more than feelings. Psychologist Derald Wing Sue defines racial microaggressions as "the everyday slights, insults, putdowns, invalidations, and offensive behaviors that people of color experience in daily interactions with generally well-intentioned White Americans who may be unaware that they have engaged in racially demeaning ways toward target groups." While microaggressions are often referred to in a racial context, anyone in a marginalized group can be subjected to them due to their age, gender, religion, ability, or sexual orientation. Consider whether you have said or done any of the following:

- **Age:** Told a young person that they're "too young" to know or have an opinion about something. Or, declined technical assistance from an older person as their age convinced you that they could not possibly be technically savvy.

- **Sexual orientation:** Refused to acknowledge the correct gender pronoun of an individual or continuously use the wrong one.
- **Religion:** Asked a Muslim woman if she was hot wearing that thing—referring to her hijab.
- **Gender:** Referred to a female co-worker in a seemingly endearing term, i.e., sweetheart or honey.
- **Ability:** Complimented a person with a physical disability for doing ordinary things, i.e., opening a door.
- **Race:** Called authorities to report a Black man walking through the office or told an Asian person, "Your English is really good."

Whether intentional or not, consciously or unconsciously, the harmful effects are cumulative. Researchers have characterized experiencing frequent microaggressions as death by a thousand cuts, and over time they can take a toll on a person's well-being. For years, I believed that I was being too sensitive and that I needed to develop a thicker skin. I made it a habit to disregard or minimize them. I didn't see that I had any other choice—it was just so commonplace. After all, what words would I choose in telling my boss that I felt weird each time he referred to me as "sweetie" or male co-workers who focused on my bust rather than my words during conversation? How was I to process the barrage of inquiries after the company picnic on whether my date, who drove a Mercedes Benz, was a drug dealer? I was ignorant about how to articulate what was going on and how it was making me feel. I never understood that that weird feeling was my body reacting and my mind telling it to chill out. Now that I know more, I realize that it wasn't me being sensitive. It was more than hurt feelings; it was stress triggers. They were a constant reminder that "you are not one of us," repeatedly dismissed, invalidated, insulted, or alienated for being an African American female and, clearly, not one of them.

American Psychological Association reports:

In addition to being communicated on an interpersonal level through verbal and nonverbal means, microaggressions may also be delivered environmentally through social media, educational curriculum, TV programs, mascots, monuments, and other offensive symbols. Scholars conclude that the totality of environmental microaggressions experienced by people of color can create a hostile and invalidating societal climate in employment, education, and health care.

The report also points out that critics downplay and minimize the impact of microaggressions regarding them as trivial and insignificant offenses that are no different from common incivilities and everyday rudeness. Further, they assert that people of color are being taught to catastrophize microaggressions and be intolerant of offenses. Psychologist Sue distinguishes how racial microaggressions differ from "everyday rudeness" in that they are:

- Constant and continual in the lives of people of color
- Cumulative in nature and represent a lifelong burden of stress
- Continuous reminders of the target group's second-class status in society
- Symbolic of past governmental injustices directed toward people of color (enslavement of Black people, incarceration of Japanese Americans, and appropriating land from Native Americans)

While there are those in the workplace who agree with the critics, my personal experience contradicts those beliefs. The American Psychological Association Report goes on to share that the burden of contending with a lifetime of microaggressions have been found to:

- Increase stress in the lives of people of color
- Deny or negate their racialized experiences
- Lower emotional well-being
- Increase depression and negative feelings
- Assail the mental health of recipients
- Create a hostile and invalidating campus and work climate
- Impede learning and problem solving
- Impair employee performance
- Take a heavy toll on the physical well-being of targets

As we march toward more just and equitable workplaces and societies, we have a choice to make—are we going to continue to perpetuate the status quo or actively work to change it? Activity 8, "Slow Down Before You React," discusses how to respond upon discovery that a microaggression has been committed and someone was offended. Diminishing the effects of microaggressions requires focus on the harmed or offended individual and contributes to inclusion and belonging. Deciding on whether to take action is inevitable—whether calling out the aggressor or supporting the victim. Ideally, we need to be prepared to do both. Review Activity 26, "If You See Something, Say Something," for details. Most of us never consider the impact of microaggressions and leave people to process and heal on their own. Some may prefer it that way, so it's best not to rush in to be the hero. Rather, have a plan to know what to say and how to behave. Begin by letting them know that you noticed what happened and ask if they are alright. Ask whether they would be okay with you doing or saying something. If they are open to support, they may offer recommendations or even ask what you think can be done. Be prepared with ideas and keep any promises made. As the effects can be long term and become masked over time, consider checking in again a few weeks later. Champions of inclusion recognize microaggressions, understand their impact, and work to educate the offender while supporting the offended.

Actions

Facilitate Circles of Trust

Imagine a safe space (virtually, in-person, or hybrid) with a small group where people can share their experiences openly and honestly without judgment or repercussion. A space where everyone listens to and supports one another. Invalidations of how people feel are not allowed as everyone is the expert on their feelings. What happens in the space stays in the space. Circles of trust are especially beneficial in workplaces that don't have affinity groups in place. Getting together regularly strengthens relationships as folks share and learn from one another. A *New York Times* post shares that:

> **Shardé M. Davis, a professor of communication at the University of Connecticut, has studied supportive communication about microaggressions among groups of black women and finds that talking can facilitate the coping process. Although Dr. Davis's study was limited to black women, she believes the spirit of what that represents could easily translate to other groups of people.**

Here are resources:

- **"Creating Safe Spaces for Victims of Microaggressions"**: www .natcom.org/communication-currents/creating-safe-spaces- victims-microaggressions

- Shardé M. Davis's guide available for purchase: www.tandfonline .com/doi/full/10.1080/03637751.2018.1548769

Understand the Real Consequences of Microaggressions

Caring and empathy begins with understanding how individuals are affected. According to a post in the *New York Times*:

Discrimination—no matter how subtle—has consequences. In 2017, the Center for Health Journalism explained that racism and microaggressions lead to worse health and pointed out that discrimination can negatively influence everything from a target person's eating habits to his or her trust in their physician, and trigger symptoms of trauma. A 2014 study of 405 young adults of color even found that experiencing microaggressions can lead to suicidal thoughts.

Diminishing the effects of microaggressions requires a two-pronged approach: 1) educating ourselves and others on what they are and being intentional to avoid them, and 2) supporting individuals who have been harmed.

Develop a Habit of Checking In

Checking in does not have to occur only after an incident. Establishing and maintaining a trusting relationship with individuals from marginalized groups makes checking in less awkward. More than likely, if trust has not been established, checking in on folks is pointless. There's an increased likelihood of not getting an honest response. Developing the habit of sincerely asking people "how they feel" or "how things are going" and taking the time to listen may lead to deeper conversation and shows that we care. Become a safe and trusted space as an ally to collaborate on what's needed for a better workplace experience.

Practice Microaffirmations

You guessed it. Microaffirmations are the antithesis of microaggressions. A post on Cornell.edu describes this is as "where people of color consciously affirm each other's value, integrity, and shared humanity." I propose that we not be exclusive on this one. This is something that everybody can and should do.

Action Accelerators

- **BBC.com**: "How Microaggressions Cause Lasting Pain," by Bryan Lufkin: www.bbc.com/worklife/article/20180406-the-tiny-ways-youre-offensive---and-you-dont-even-know-it

- **HBR.org**: "The Surprising Power of Simply Asking Coworkers How They're Doing," by Karyn Twaronite: hbr.org/2019/02/the-surprising-power-of-simply-asking-coworkers-how-theyre-doing

- **Africa.BusinessInsider.com**: "Memorize these scripts so you can call out microaggressions at work and support your colleagues": africa.businessinsider.com/careers/memorize-these-scripts-so-you-can-call-out-microaggressions-at-work-and-support-your/tekm9e1

- **Today.Duke.edu**: "The Harmful Effects of Microaggressions, by Jack Frederick": today.duke.edu/2022/05/harmful-effects-microaggressions

- **American Psychological Association Disarming Racial Microaggressions**: "Microintervention Strategies for Targets, White Allies, and Bystanders": engineering.purdue.edu/Engr/People/faculty-retention-success/Files/Racial-Microaggressions.pdf

Sources Cited

American Psychological Association Disarming Racial Microaggressions. "Microintervention Strategies for Targets, White Allies, and Bystanders," engineering.purdue.edu/Engr/People/faculty-retention-success/Files/Racial-Microaggressions.pdf

Hahna Yoon. "How to Respond to Microaggressions," NYT.com (subscription required), March 3, 2020, www.nytimes.com/2020/03/03/smarter-living/how-to-respond-to-microaggressions.html

Cornell University. "The Detrimental Effects of Microaggressions," evidencebasedliving.human.cornell.edu/2021/10/12/the-detrimental-effects-of-microaggressions

47

Affect Work-Life Balance for All

"Don't confuse having a career with having a life."

—Hillary Clinton

***

The 40-hour workweek and workplace practices that exist today were created by the dominant culture in the 19th century. The last 200 years have most notably brought about a more diverse workforce that is demanding more from employers. Nonetheless, antiquated work-life expectations persist as organizational leaders expect staff to prioritize the needs of the business above all else. We have co-workers whom we see each day who exist in a perpetual state of exhaustion. There is just not enough time in the day to effectively manage the demands of our personal and professional lives to create a healthy work-life balance. Who wouldn't want to have enough time and energy left at the end of the workday to enjoy life's pleasures? While I can't think of a single reason that someone wouldn't respond with "Please, sign me up!" the reality is that many folks leave the workplace exhausted—mentally, physically, and emotionally. In some cases, exhaustion is "always on" as just the thought of work is exasperating. After putting in 8, 10, or more hours per day to earn a paycheck, there is little energy or time left to enjoy the fruits of our labor or relax and replenish. We're always "on" and can't find the "off" button—perhaps because it does not exist. With 24/7 connectivity, the pressure to excel, to always be accessible while maintaining productivity, can be overwhelming. The National Safety Council reports that "97 percent of workers have at least one workplace fatigue risk factor and more than 80 percent have two or more. When multiple risk factors are present, the potential for injuries on the job increases. Fatigue can have detrimental effects on a person's health and safety both on and off the job." A Deloitte survey reveals the following:

---

77 percent of respondents say they have experienced employee burnout at their current job, with more than half citing more than one occurrence. The survey also uncovered that employers may be missing the mark when it comes to developing well-being programs that their employees find valuable to address stress in the workplace.

Many companies may not be doing enough to minimize burnout: Nearly 70 percent of professionals feel their employers are not doing enough to prevent or alleviate burnout within their organization. 21 percent of respondents say their company does not offer any programs or initiatives to prevent or alleviate burnout.

Companies should consider workplace culture, not just well-being programs: One in four professionals say they never or rarely take all of their vacation days. The top driver of burnout cited in the survey is lack of support or recognition from leadership, indicating the important role that leaders play in setting the tone.

Achieving balance or managing work with life responsibilities is pretty much impossible without appropriate organizational policies and benefits as well as the support of leaders and colleagues alike. The life part of work-life balance gets crowded out when we take work home, are expected to respond to email and instant messages after hours, or are called in to work on a scheduled day off. While many employers now have policies in support of work-life balance like flexible hours, remote or hybrid models, generous paid time off, and meeting black-out days, staff may be hesitant to take full advantage of them. The reasons vary by individual and work climate. These are the ones I hear most from friends and colleagues:

- Concerns over the work piling up—there's never a good time to take off more than a day or two
- Staying viable and alert in highly competitive environments

- Returning to an overloaded email inbox and feeling pressured to catch up promptly
- Projects taking a turn for the worse in their absence
- Fear of placing their careers in jeopardy
- Necessity for overtime pay
- Pressure to accept additional shifts
- Labeled "not a team player" for prioritizing non-work-related plans

Some may be convinced that better time or money management is the solution. The fact of the matter is that many members of marginalized groups don't earn a living wage to begin with, so there's constant struggle to make ends meet. Thus, extra shifts and overtime pay are welcomed. Members in marginalized communities who do earn a living wage often experience greater challenges in achieving balance due to bias and stereotypical beliefs as well as systemic inequities blocking access to much needed resources. In addition, when work responsibilities bleed into personal time, it's beyond one's control when you're an essential worker or individual contributor. With so many knowledge workers in back-to-back meetings all week, the only time to get any actual work done is before or after the workday. The Great Resignation and organizations finding themselves shorthanded on frontline workers adds extra burden to the ones who stayed. Despite current conditions, the common assumption persists that those seeking more balance are less driven or less committed to the organization and that remote workers require micromanaging or are "getting away" with something. These beliefs have a way of cropping up in day-to-day interactions and can inadvertently discourage co-workers from taking advantage of policies that support work-life balance. When a colleague learned that I would be vacationing for two weeks, his initial reaction wasn't what I expected. I was accustomed to responses that show excitement or interest and an assumption of fun or restful activities while away. His response was, "Why so long? Don't expect me to cover your accounts while you're out. I've got enough going on." I did not take it personally. He was clearly more concerned about any extra work that may come his way—and I can appreciate his point of view having covered for many colleagues myself. However, it would have been nice to have his support rather than leaving with apprehension about what may be in store for me upon my return.

Willow, a project manager in a hybrid environment and single mother of two grade-school children, looks forward to "No Meeting Mondays" at her firm. She finds that Sundays are less stressful in not having to prepare for a barrage of meetings and is more productive on Mondays. She can approach the day with a clear head and a clean slate while the kids attend school. The policy looks good on paper. Unfortunately for her, while there are no actual meetings, the constant interruption of impromptu calls that may last 30 minutes or more feel like unplanned, unscheduled meetings. Juan, a middle-aged factory worker with chronic lower back pain, spends a lot of time on his feet. Though he is approved for Family and Medical Leave Act (FMLA) benefits, he does not take advantage as often as he would like. He considers co-workers family and is keenly aware of the disruption to their lives when they work extended hours in his absence. Some make him feel like he is letting them down and not carrying his weight when he overhears their complaints about crew members continuously taking off. The balancing act can become difficult especially for colleagues who shoulder the responsibility of caring for children and/or aging parents, as well as those who have identities outside work, i.e., Little League coach or Bible study teacher. The struggle is no different for those who need less time in the "fight or flight" zone and more time feeling centered and calm, while others need more time just to think and create. Balance is not about working less or getting by with minimum effort; it's more about being in an environment that supports one's ability to be the best version of themselves in and outside the workplace. No one expects a 50-50 balance between their professional and personal life but more so the capability to be in more control of the two.

Work-life policies are more easily implemented when we are supported by managers; however, we must feel supported by co-workers as well. It's our day-to-day interactions that empower colleagues to use policies and build a work-life model that supports their needs and still achieve maximum productivity. Creating an inclusive workplace means looking at balance through the lens of inclusion where all employees can thrive in their unique situations. Everyone plays a role. Sending an email after hours? Let the recipient know that it's okay to respond on the next business day. Before placing that phone call, first send a message checking whether now is a good time to talk. While using the "scheduling assistant" feature for calendar invites may be productive for you, it may be disruptive to your colleague when they suddenly realize there's an unexpected meeting to plan for and attend. What harm would it do to ask for the best time to connect? While we don't expect the status quo to totally disappear, workplaces are more inclusive of all when the status quo includes balance. Champions of inclusion take note of signals that a co-worker's balance may be off and are mindful to check in and offer support. It's a great feeling to know that you've made a difference in someone's day.

Actions

Establish New Personal Norms

Everybody has a workstyle that works best for them and their circumstances and may take advantage of company perks like half-day Fridays, flexible start and end times, and the option to work from home. We can be respectful of all situations when we create norms that are accommodating regardless of unique circumstances. Consider adding these norms to your day to day:

- Add a statement to your email signature like, "If you are receiving this outside of working hours, there is no expectation to respond until you return."
- When sending virtual meeting invitations, include "camera is optional."
- Respect blocked time on co-workers' calendars by not scheduling meetings with them during that time.
- Show respect for co-workers' workload by not making last-minute urgent requests.
- Give people the benefit of the doubt when co-workers unexpectedly miss work.
- Give grace. When people take extended vacations, it's only natural that the workload will be heavier not only for those who are charged with covering the work in the absence of a teammate but for that individual upon their return. Be willing to extend a deadline or give a friendly reminder or follow-up. When folks are overwhelmed, things slip through the cracks.
- Show appreciation for those who cover in your absence. Send a thank-you note and compliment them on a job well done—especially if they went above and beyond. While showing appreciation won't alter the workload, the gratitude can increase feelings of happiness.

Make a Collective Effort

Collaborate with team members on how they can best be supported in achieving balance to create an all-for-one and one-for-all team environment. Things work better when we work together. Hearing it directly "from the horse's mouth" removes assumptions and positions us to be more mindful of unique situations. Work together to create "team best practices for balance." Revisit it regularly as circumstances change. Share the idea with other teams. Involving your manager in this discussion or sharing the plan after it has been created may be prudent. Remember, the process works only when it is supported by both manager and individual contributor.

Look for Warning Signs

As relationships are forged within the workplace, co-workers will "tell" you that they may be struggling to achieve a healthy work-life balance without uttering a word. When we're accustomed to seeing folks in a healthy state of being, it's easier to spot signals of distress. Typical signals include inability to focus or stay alert and being short-tempered, less talkative, or distant. Professional and personal lives can suffer as folks battle to stay afloat. Professionally, there is an increased likelihood of mistakes being made and missing deadlines or important details. Personally, they may miss out on important life events and avoid socializing in attempts to manage work overload. When stress signals become obvious, do the following:

- Encourage co-workers to take a vacation day.
- Remind them of the importance of taking breaks and lunch.
- Encourage them to take nonworking lunches and breaks.
- Alert them that you've taken note of their changed demeanor and ask whether there is something you can do to help.

Action Accelerators

- HBR.org: "How Companies Make It Harder for Lesbian, Gay, and Bisexual Employees to Achieve Work-Life Balance," by Katina B. Sawyer, Christian N. Thoroughgood, and Jamie J. Ladge: hbr .org/2018/08/how-companies-make-it-harder-for-lesbian-gay-and-bisexual-employees-to-achieve-work-life-balance

- **HBR.org**: "The Surprising Benefits of Work/Life Support," by Alexandra Kalev and Frank Dobbin: hbr.org/2022/09/the-surprising-benefits-of-work-life-support

- **SHRM.org**: "How the Term 'Work-Life Balance' Is Changing for the Youngest Group of Workers," by Tracey Brower: www.shrm.org/resourcesandtools/hr-topics/people-managers/pages/work-life-balance-for-youngest.aspx

- **Inclusion-inc.com**: "Taking an Inclusive Approach to Work-Life Balance": www.inclusion-inc.com/single-post/taking-an-inclusive-approach-to-work-life-balance

- **BookGraham.com**: "It's not just about 'balance' anymore: Work-life inclusion": www.brookgraham.com/diversity-and-inclusion-blog/its-not-just-about-balance-anymore-work-life-inclusion

Sources Cited

Deloitte.com. Workplace Burnout Survey, www2.deloitte.com/us/en/pages/about-deloitte/articles/burnout-survey.html?id=us:2sm:3li:4dcom_share:5awa:6dcom:about_deloitte

National Safety Council. NSC Fatigue Reports, www.nsc.org/workplace/safety-topics/fatigue/fatigue-reports

Value People *Because* of Difference—Not in Spite of

"It is not our differences that divide us. It is our inability to recognize, accept, and celebrate those differences."

—Audre Lorde

***

Inclusion is everybody's job, and it's a job that anybody can do. Every interaction with our co-workers makes them feel either excluded or included. We have practiced inclusion and exclusion since we were children while learning to play with other kids. Over time, we developed preferences for playmates. Those we liked were usually extensions of ourselves, so we continued to form relationships with them (included); and those we liked less or not all, we avoided, and no relationship was formed (excluded). Back then, we were not consciously including or excluding; it was simply about preference. As we matured and socialized with more groups, we became more intentional about who we established relationships with. Often, included individuals were similar to us and shared common interests—and differences such as race, culture, age, etc., may have been less important unless, of course, you were raised to believe that they were. For the most part, we had a group that we identified with and felt safe to be ourselves around—shutting out all others. When an outsider wants to become a part of the group, they must prove themselves to not be a threat to group norms, learn those norms, and demonstrate that they have a similar mindset to be allowed entry. This dynamic has followed us into the workplace.

As the faces in the workforce become more diverse, our differences are obvious and are often unwelcomed, so inclusion remains hard to come by. People who are different from the dominant culture are expected and/or pressured to blend in or assimilate to maintain the harmony and get the work done as usual. The reason that organizations are diversifying their workforces isn't for diversity's sake or even because it's the right thing to do (although it is); it's for the proven value that diversity brings. One study after another validates that companies that figure out how to successfully leverage a diverse workforce gain a competitive advantage over those with homogenous workforces. The McKinsey Report "Diversity Wins: How Inclusion Matters" states that "Diverse teams are more innovative—stronger at anticipating shifts in consumer needs and consumption patterns that make new products and services possible, potentially generating a competitive edge." The report further expands on the impact of race and gender diversity:

> Moreover, we found that the greater the representation, the higher the likelihood of outperformance. Companies with more than 30 percent women executives were more likely to outperform companies where this percentage ranged from 10 to 30, and in turn these companies were more likely to outperform those with even fewer women executives, or none at all. A substantial differential likelihood of outperformance—48 percent—separates the most from the least gender-diverse companies. In the case of ethnic and cultural diversity, our business-case findings are equally compelling: in 2019, top-quartile companies outperformed those in the fourth one by 36 percent in profitability, slightly up from 33 percent in 2017 and 35 percent in 2014. As we have previously found, the likelihood of outperformance continues to be higher for diversity in ethnicity than for gender.

The promises of a diverse workforce can't be realized until we embrace different perspectives in how the work gets done. Early in my sales career, I was the only African American woman on a sales team of 20 individuals at a technology services and support company. The organization had a sales methodology that consisted of making 100 cold calls a day. It was a churn-and-burn environment—get to the decision-maker, share your value proposition, get the appointment, close the sale. The weird thing was that while the goal was to make a sale, the metric that mattered most to them was the 100 cold calls. After 30 days of that with almost nothing to show for it, I was convinced that something had to change. I had recently read a book on relationship selling and spent time in the evenings to integrate those practices into my sales strategy. Yup. I put in more hours, and my sales started to improve. Building relationships is something that comes natural to me, and it paid off. By the sixth month, I was consistently achieving quota. My approach to selling was different, yet very effective. Once we accept the fact that just because something is different does not render it wrong, less valuable, or inappropriate, transformation can begin. When I started a new role that required me to facilitate monthly meetings, I was the only African American on a team of five, and one of a few in an organization of two hundred. My boss's process was to lead discussions using PowerPoint to drive the interaction, and that's the methodology that my peers were using as well. Having to create and adjust slides to be cohesive in real time while building my skills to be an effective facilitator drove me nuts. I decided I could be a better facilitator if I had a template that was prepopulated with topics and ideas that could be evaluated and expanded upon during sessions. I built it in Word. Not only was it easier for me to manage, but I could put more energy into interacting with participants, which yielded more engagement—and we were able to get a lot more accomplished in our time together. When my boss showed up unannounced, I had a brief panic attack that I would be chastised for changing the process. Afterward, he shared that he loved the energy, said how well I was doing with the new format, and congratulated me.

Diversity is less about how a person looks and where they come from and more about valuing different points of view and being open to new ways of thinking. It's about integrating diversity of thought, which can come only from the unique backgrounds and experiences of diverse individuals. *Psychology Today* offers more perspective on homogeneous teams: "Dozens of studies and decades of research have found that diverse teams tend to be smarter than homogeneous teams: they often think more logically, are more creative, and are more adept at identifying errors in thinking." Organizational leaders continue to make missteps along racial lines when the status quo goes unchallenged. When NPR's Ari Shapiro interviewed race and culture reporter Nikole Hannah-Jones of *The New York Times Magazine*, they discussed racially insensitive ads by Ancestry, Nivea, Ram Trucks, and Gucci, and why they keep happening. Hannah-Jones explains:

> We have a very unsophisticated way of understanding race and racism and the history of race and racism in this country. Many people don't really understand, particularly white Americans, that the language that they're using and the depictions that they're coming up for our advertising are offensive. I think another reason is, a lot of times, the teams who are coming up with this advertising don't have people of color on the teams or at least people of color who can have a strong, influential role on the campaigns that come out.

I agree with her. My experience of being in the minority at most every place I've ever worked resonates with her assessment. She goes on to share in reference to Ancestry:

> I heard from a couple of folks, but no one who worked directly on this campaign. What I did hear is more generally about the problems in the ad agencies—that often they are not very diverse and that if there are people of color who serve on these teams, they do not feel empowered enough to push back on some of these campaigns.

Diverse representation must be empowered through inclusion to fully participate, which will not only render a fresh perspective but can help avoid catastrophic mistakes and insensitivities as referenced by NPR. Our role is to adjust our thinking and view difference as a strength. It's an opportunity to do things better. Each individual brings something of relevance and value through their lived experiences to problem solving, creating, managing through crises, and other business scenarios that improve business outcomes. We must set the table where all voices can be heard and considered. Champions of inclusion make it a habit to tap into the expertise of diverse team members and endeavor toward making it a team-wide effort. When more people are consistently intentional in unleashing inclusion, the value of difference can emerge.

Actions

Get to Know Co-workers and the Skills They Bring

Get to know co-workers' capabilities, what they have done in previous roles, and what they do better than anyone else. When we know who has what skills, the better they can be leveraged as well as complement the skills of others. The more we know about what they know, we can open the door to more meaningful collaborations on shaping goals and processes or driving direction and strategy. Get to co-workers as people—not just as co-workers. Create networks or events around common interests like pets or cycling, for example (See Figure 48.1). Review Activity 4, "Make the Connection;" Activity 12, "Strive to See the Whole Person;" and Activity 25, "Foster an Environment of Trust."

Figure 48.1: Collaborate across differences.

Be Mindful of Habits That Exclude

So often people are interrupted or ignored, and we miss out on their contributions. Disrupt any actions that prohibit full participation. The next time you want to "pick someone's brain" or "bounce an idea" past someone, instead of reaching out to the usual suspects, reach out to someone who is different from you. Remember, enabling difference transforms how the work gets done, and we must be open to new possibilities. Whatever you are working on right now, consider what competencies or voices may be lacking that may enhance the outcome, and go treasure hunting in diverse pools.

Seek Similarities

Real inclusion means not being made to feel the need to hide one's true self. The energy it takes to accomplish that day in and day out takes away from the energy needed to do one's best work. It is bias and intolerance of our differences that divides us. Our differences in physical attributes and lived experiences will never change; thus, we must move past tolerating difference to valuing it. Rather than focusing only on difference, look for commonalities and ways to unite. For example, I value hard work, ethics, integrity, and critical thinking, and I gravitate toward individuals who demonstrate those values regardless of how different we may be. That signals to me that we both care about high-quality work and will do our best to ensure that it is. I befriended a younger, White, male colleague when I overheard him talking to his financial planner. I was intrigued and wanted to learn from him, and he was only too happy to help. Our small talk always included something new he had learned that may benefit me. We connected over financial security. In conversation, listen carefully for clues. (For more on listening, review Activity 22, "Amplify Voices That Aren't Being Heard.") Perhaps you grew up in the same hometown and can connect over similar experiences. Maybe you'll learn that you share an invisible disability and can now support one another. Who knows what you'll discover? This will take practice and requires intention with a curious mindset (See Figure 48.2).

Figure 48.2: Build bridges to identify common interests.

Action Accelerators

- **Yahoo Finance**: "The Subtle Ways People of Color Feel Excluded at Work," by Ellen McGirt: finance.yahoo.com/news/subtle-ways-people-color-feel-190600682.html?fr=sycsrp_catchall

- **Catalyst.org**: "The Day-To-Day Experiences of Workplace Inclusion and Exclusion," by Julie S. Nugent, Alexandra Pollack, and Dnika J. Travis: www.catalyst.org/wp-content/uploads/2019/01/the_day_to_day_experiences_of_workplace_inclusion_and_exclusion.pdf (Note: This resource is geared toward leaders but provides insight for everyone else.)

- **YouTube.com**: *The Power of Exclusion*, Tiffany Yu, TEDxBethesda: www.youtube.com/watch?v=qVtDejw8ZBw

- **TinyBuddah.com**: "When You Don't Fit In: The Value of Being Different," by Devi Clark: tinybuddha.com/blog/value-of-being-different/#:~:text=Being%20different%20has%20tremendous%20value.%20Here%20is%20how.,gay%20people%2C%20introverts%2C%20recovering%20addicts%2C%20and%20many%20others

Sources Cited

McKinsey.com. Diversity Wins: How Inclusion Matters, May 19, 2020, www.mckinsey.com/featured-insights/diversity-and-inclusion/diversity-wins-how-inclusion-matters

David Rock. PsychologyToday.com: Why Diverse Teams Outperform Homogeneous Teams, June 04, 2021, www.psychologytoday.com/us/blog/your-brain-work/202106/why-diverse-teams-outperform-homogeneous-teams

NPR.org. Companies Continue to Stumble Over Racially Offensive Advertising Campaigns, April 22, 2019, www.npr.org/2019/04/22/716096440/companies-continue-to-stumble-over-racially-offensive-advertising-campaigns

Evolve from "Not Racist" to Antiracist

"In a racist society it is not enough to be non-racist; we must be anti-racist."

—Angela Davis

***

While racism in America is centuries old, the willingness to talk about and address it is very new. Individuals and organizations alike are embarking upon courageous conversations about race as we strive toward more equitable and inclusive workplaces and societies. A growing number of Americans now acknowledge the racial injustice and discrimination experienced by people of color and want to demonstrate that they are not contributing to the problem. In doing so, they are quick to point out that they are not racist. Racists actively discriminate and inflict harm based on the belief that their race is superior to another, and that's not who they are. To be clear on what is meant by racism, the National League of Cities describes it as:

A system of oppression that is based in and upholds the superiority of White people and the inferiority of Black, Indigenous, and People of Color. Racism was created by and

upheld through policies, practices, and procedures to create
inequities between racial groups. Current leaders in positions
of authority in government, private, and non-profit institutions
have inherited and upheld racist institutions and structures. These
leaders have historically diluted the term "racism" to reflect only
overt interpersonal racist actions so that the full system of racism
can continue to grow and persist, unseen and unchallenged.

So, what do people mean when they proclaim to be "not racist"? It's a
relative term. For some, not racist means that they are not a member of
the Ku Klux Klan, never owned a slave, and don't say the "N" word. For
others it means "I would never do anything to harm a Black person." And
then there are those whose stance is "I like Black people—so it's impossible for me to be racist." Some believe that racism no longer exists. Their
beliefs are validated with comments like "I voted for Obama—twice,"
"I have a good friend who is Black," or "We're all human—I don't see
color." Though many White people are aware of how differently they
are treated in day-to-day situations compared to people of color, they
are oblivious to the daily indignities that are inflicted and the systems
of oppression of which they derive great benefit. Asserting to be "not
racist" is easy when one has never experienced racism in the ways that
Black and Brown people have. Further, it does not take into account the
historically racist views that justify the unfair treatment and oppression
of people of color. To be "not racist" is a passive response to ending racism. Many contribute to the problem without realizing it and continue
to perpetuate systems of inequity in daily interactions. To be "not racist"
requires no action. It's like being a crossing guard and not pushing a
pedestrian out of the way of the oncoming bus and then justifying the
inaction by saying that they weren't the one who was driving the bus.
When one is not actively addressing a problem, they remain part of the
problem. No action, no progress. Solutions require taking action, and the
solution to ending racism requires actively participating in the creation
of justice and equality for all—in other words, being antiracist. Action
is what makes the difference between being "not racist" and antiracist.

No one is born racist any more than being born antiracist. What makes
us one or the other lies in the choices we make. Antiracism is demonstrated
through the conscious decision to make frequent and consistent choices
that lead to equitable outcomes. It's more about what we do—not who
we are. It's painfully clear to people of color how race influences and
shapes our lives—right down to the details. Black women must consider
their hairstyle before entering the workplace at the risk of being deemed

unprofessional. Black men must consider assertive mannerisms at the risk of being perceived as aggressive or violent. Asian women must be mindful about speaking up and saying what they want to avoid being perceived as weak or docile at the risk of being invisible or bulldozed. These seemingly minor details need never enter the minds of members of the dominant culture as they go about their day. In the absence of antiracist choices, people consciously or unconsciously uphold aspects of white privilege, white supremacy, and unfair systems. Evolving from "not racist" to antiracist demands a raised level of consciousness about how one perpetuates racism followed by a foundational shift in attitudes and behaviors. It will require an education about the real American history as opposed to whitewashed versions and about how racist ideas have been standardized in social systems, media, and culture. We must challenge underlying thoughts and assumptions about racism and become more intentional about achieving equitable outcomes in all that we do. Champions of inclusion are change agents toward equitable outcomes through ongoing learning, consistent reflection, and more importantly altering old habits and beliefs.

Perry, a white male in his mid-50s, considers himself an ally to historically marginalized co-workers. He has come to recognize inequities over the years, aspires to drive change, and considers himself part of the "woke" culture. He's been developing new relationships to build cultural competence and better connect with diverse co-workers while working to understand the issues they face. He readily admits that the journey has not been easy—especially when it comes to building trust and overcoming the fear of getting it wrong. Facing his biases and stereotypical beliefs has been a humbling and uncomfortable journey of self-discovery. His evolution from "not racist" to antiracist began when his daughter married a Black man and gave birth to a biracial son—his first grandchild. The thought that his grandson could be the next Trayvon Martin, Ahmaud Arbery, or George Floyd horrified him. In this moment, he realized that he could be a better ally to people of color. He was now forced to dive deeper into his long-held beliefs and attitudes specifically around race and how they may impact the relationship with his grandson, son-in-law, and extended family. To better understand the realities of race and being Black, he joined the Black-focused affinity group in his organization. There, he was able to learn directly about their experiences, both inside and outside the organization. In listening to their stories and the impact on their lives, he felt comfortable sharing his story and how he wanted to become more sensitive to personal racial bias and unintended racist behaviors. He actively took on tasks in support of group goals and learned more from them about leveraging privilege. As relationships forged, he began to mentor a few members in the group and soon realized how much he was growing in cultural competence and improving in cross-cultural communications. Accepting vulnerability and not becoming defensive when mistakes were made was critical to making progress—not only with co-workers but family members as well. He valued the feedback of mentees that helped him understand the impact of unintended offenses. It really mattered to him to know whether he was doing more harm than good.

There are many actions that we can take to promote equity over time. For instance:

- Speaking up when you hear a racist joke or witness discrimination— review Activity 26, "If You See Something, Say Something."
- Understanding and leveraging privilege—review Activity 32, "Leverage Your Privilege."
- Investing in the future of a person of color—review Activity 42, "Pick a Ship and Get on It."

Keep in mind that some actions simply raise awareness and demonstrate one's personal level of awareness like marching in Black Lives Matter protests, putting up yard signs, or sharing a social media post about racial injustice. In addition, think twice before commenting on someone's race. Consider whether it's actually relevant to the conversation. Someone once told me that I was smart for a Black girl. Why couldn't I just be smart? What did my race have to do with it? The unintended insult was that Black people are not smart or are intellectually inferior. At the end of the day, it's all about creating inclusive environments with equitable outcomes. If actions are not creating inclusive environments with equitable outcomes, it's best not to employ them.

Actions

Reflect and Decide

The following videos represent acts of racism. As you review them, consider the consequences of being "not racist" and antiracist. Place yourself at the scene and reflect on how "not racist" attitudes may or may not maintain, protect, or contribute to the circumstances. In your opinion, do you believe feelings of entitlement influenced the offenders' actions? What do you believe to be the root causes of their actions? Take note of your own attitudes about race and how they informed your responses. Ask yourself how an antiracist would respond to create an equitable outcome and compare that to "not racist" responses. How much do they differ? Decide if you are willing to respond in the same way as the antiracist. Why or why not?

- **YouTube.com:** *Central Park Karen* (full video), www.youtube.com/ watch?v=aPvb_mszKew

- **YouTube.com:** *Hotel Employees Fired for Calling Cops on Black Guest,* www.youtube.com/watch?v=XqnNV_Ku9Oo&t=85s

- **YouTube.com:** *Woman who attacked teen, falsely accused him of stealing her iPhone in hotel identified,* www.youtube.com/watch? v=Un6zVW-K0qQ

- **YouTube.com:** *NJ Man Accused of Racial Harassment to Face More Charges After Protests Outside His Home,* www.youtube.com/watch? v=4su9fF802Yc&t=158s

- **YouTube.com:** 2 *Black Men Arrested At Philadelphia Starbucks Receive $1 Settlements,* TODAY, www.youtube.com/watch?v=D0Aj08cntf4

- **YouTube.com:** *High school students seen in blackface spark protests near Chicago,* www.youtube.com/watch?v=1zXYYLDLMUU

- **YouTube.com:** *Woman fired after blocking black man from entering his apartment building,* www.youtube.com/watch?v=k7sNKxvaEi4

- **YouTube.com:** *BBC Ideas: How I deal with microaggressions at work,* www.youtube.com/watch?v=fzyL96LZdHU

Examine Your Beliefs About Race

Antiracism focuses more on what we do that perpetuates racism and less on who is racist. Many people recognize overt racism as seen in the videos in the previous exercise. However, even small displays of racism contribute to keeping systematic racism in place—like telling a Latina that she is a credit to her race. The unintended insult was a blatant generalization that LatinX people are not good people, but the one in front of you is acceptable. Get in touch with your thoughts and attitudes about race and commit to continuous self-examination. Consider actions to take or old habits to break to evolve from "not racist" to antiracist (see Figure 49.1). Think about the last time you did anything in support of people of color and how often you do so. If you are having difficulty remembering, there's work to do. Consider whether the belief that one is "not racist" absolves them of the responsibility to do better.

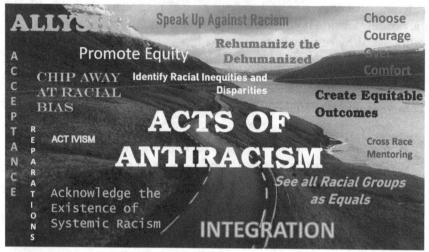

Figure 49.1: Acts of antiracism

Action Accelerators

- ■ CBSnews.com: *The difference between being not racist and being anti-racist,* by Christina Capatides: www.cbsnews.com/news/antiracist-not-racist-difference

- ■ YouTube.com: *"but I'm not racist?"* UnJaded Jade: www.youtube.com/watch?v=NTmNV5KTdDo

- ■ YouTube.com: *The difference between being "not racist" and antiracist,* Ibram X. Kendi: www.youtube.com/watch?v=KCxbl5QgFZw

- ■ YouTube.com: *10 Ways to Promote Anti-Racism In The Workplace,* Forbes: www.youtube.com/watch?v=fezoP7QbIkY

- ■ NPR.org: *Work to be Antiracist:* www.npr.org/2020/08/24/905515398/not-racist-is-not-enough-putting-in-the-work-to-be-anti-racist

Sources Cited

NLC.org. What Does It Mean to Be an Anti-Racist? www.nlc.org/ article/2020/07/21/what-does-it-mean-to-be-an-anti-racist - :~:text="Not racist" often refers to a passive response, has operated in this country since its "founding."

Create the Future by Learning from the Past

"Those who don't know history are doomed to repeat it."

—Sir Edmund Burke

***

Our lives represent living history. We are derivatives of past experiences, human behaviors, and attitudes. Almost every aspect of our being was inherited from the past—from the language we speak to our cultural traditions to the religion we ascribe to our genetic makeup. If you have ever checked out your grandma's family photo album and realized that your smile is identical to an uncle that you never knew existed or that your body shape has a striking resemblance to your great-great-grandmother, then you've noticed your connection to history and how it transcends or even dictates the future. It's usually sheer delight when we recognize inherited personality traits and physical features, but what about the ones that cause us to struggle? We are now learning more about and discussing something else that can be inherited—intergenerational trauma. A post on Forbes.com shares how this relates to the Black community:

One major consequence of settler colonialism is the trauma that is passed down from generation to generation—what psychologists label as intergenerational trauma. The American Psychological Association defines this as "a phenomenon in which the descendants of a person who has experienced a terrifying event show adverse emotional and behavioral reactions to the event that are similar to those of the person himself or herself." For Black Americans who descended from enslaved people, the trauma that was experienced by ancestors has been passed down through each generation.

As we have continued to grapple with systems of racism and oppression for centuries, the associated trauma doesn't die with the traumatized individual. The term *weathering* was introduced in 1992 by public health researcher and professor Dr. Arline Geronimus of the University of Michigan, and according to `MedicalNewsToday.com`, her work has received more than 1,000 citations with the most substantial rise in 2020. Research conducted by Geronimus and colleagues states:

Findings provide evidence that the impact of chronic stress on health has important implications not only for individuals but also for the population as a whole and suggest ways that dynamic social relationships between racial and ethnic groups may shape health in a race-conscious society. The findings suggest that progress in understanding and eliminating racial health inequality may require paying attention to the ways that American public sentiment on race, including its gendered aspects, exacts a physical price across multiple biological systems from Blacks who engage in and cope with the stressful life conditions presented to them.

Until we start addressing and eradicating the root causes of systemic racism, people of color will continue to be stigmatized and disadvantaged for generations to come.

While American society and organizations profess to be open and egalitarian, the world in which we live today is in many ways rooted in the past—from policies and politics to systems and ideologies. People of color continue to feel acutely disrespected in everyday lives as we experience both subtle and explicit forms of discrimination. There's a reason that there are so few people of color at the helm of Fortune 500 companies and are not developed through the ranks at the same rates as White counterparts. There's a reason for the Starbucks incident that led to the arrest of two innocent black men who were simply waiting for a friend. There's a reason that the Jackson National Life Insurance Company agreed to a settlement of $20.5 million after being found guilty of discriminating against Black and female employees in hiring and promotion practices. There's a reason that the JL Schwieters Construction company of Minneapolis settled for $125,000 for creating a hostile work environment for two Black carpenters. History tells us that slavery effectively dehumanized Black people and that emancipation gave us freedom but relegated us to the bottom rungs of society, allowing many White people to continue to feel superior. To this day, racist values are passed from generation to generation while legislation and policies continue to discriminate and oppress.

Currently, we are creating history. The way we live our lives and the decisions that we make are writing the next chapter. The continuation of racism, hatred, and oppression does not have to be inevitable. We have the power within us to create the future we want to see where all can thrive or at least have an equal chance to do so. We must study the ugly and harsh realities of the past that brought us here, examine what went wrong and why, and work toward solutions that dismantle institutionalized and systemic racism. America is in the midst of a racial reckoning. People are looking in the mirror for the first time and acknowledging personal accountability for prejudiced, biased, and racist beliefs. By contrast, there are others who deny the brutal past and are working to ensure that we and future generations never learn the ugly truth. They are expunging history and replacing it with properly sanitized versions—versions that will change each time a new leader emerges and decides to change the narrative to align with their political positions. A post on the website of the American Association of Colleges and University (AAC&U.org) shares:

> **Texas is not alone in its efforts to restrict education on racism, bias, and the contributions of specific racial or ethnic groups to US history. As of mid-July (2021), twenty-five other states had introduced similar bills or other efforts targeting academic lessons, presentations, and discussions of "divisive concepts" in schools, colleges, and universities. Political jockeying aimed at controlling which history gets recognized, suppressing inconvenient truths, and resisting initiatives to tell a more comprehensive, inclusive story of our nation's history has taken hold in this post-truth environment in which facts are less influential in shaping public opinion than appeals to emotion and personal belief.**

The post continues:

But the fact is that "Education Not Indoctrination" is the rallying cry on both sides of the debate over how history should be taught. Politicians advocating for restrictive legislation are responding to a growing populist base that asserts that a liberal bias exists in higher education and exhibits increasing contempt for experts. Under the circumstances, engaged pluralism, grounded in collaborating across difference, is more critical than ever for social transformation and for the strength of our nation. Educating for democracy requires taking our messages beyond the academy and partnering with public artists and humanists in shaping an inclusive American story in ways that are accessible to all.

In erasing history, we lose context of not only what went wrong and why but what worked well and why. Erasing history removes our opportunity to learn as it provides a reflection of the past where we can compare values, behaviors, and motives. While the reflection may reveal aspects of ourselves that we're not proud of, it casts us a line to do better and be better in the future. We must stop looking at others and deciding what *they* need to do. The status quo will remain prevalent as long as we look, wait, and hope for someone else to do what we can do ourselves. Knowing what we know now, we must look in the mirror and ask, "What am I doing to create a more just and equitable society?" What we do for the good of society affects everything else—including the culture of the places where we work and learn. Marcus Tullius Cicero wrote, "To be ignorant of what occurred before you were born is to remain always a child. For what is the worth of human life, unless it is woven into the life of our ancestors by the records of history?"

Factual history gives us insight into present-day issues and helps us get to the *why* of inequities and provokes us to the *how* of changing them. We learn not only stories of ordinary people who exhibited great courage in the face of adversity and actions they took to create social justice but also stories of those who were intentional to maintain systems of oppression. Armed with the knowledge of history, we can challenge our belief systems, understand why we think what we think, and test them against our vision for the future. Champions of inclusion are truth-seekers and make the effort to acquire accurate historical knowledge from diverse thought leaders. We can no longer neglect, reject, or ignore the fact that racism and systems of inequity have existed for centuries and that people of color continue to suffer from them. Historical data and information tells the story of who we are and where we come from and has the potential to reveal where we are headed. To create more inclusive cultures, we must decide how we'll use history "lessons" to inform present decisions as well as those of the future.

Actions

Launch a Truth Investigation

We are constantly inundated with information, and it's everywhere—from social media to television and news shows to books, podcasts, and webinars to music and other art forms. All of it is designed to inform, evoke an emotion, and ultimately encourage us to act or think a certain way. With so much coming at us, it's convenient to take the easy route of delving into what aligns with our current beliefs. To acquire knowledge of historical truth we must take the time to fact check details from multiple and diverse sources. This could be as simple as talking to grandparents and great-grandparents about their experiences, unearthing printed news stories from the local library or newspapers.com (online newspaper archive), visiting .gov websites to review historical documents, or learning from historians with objective versus subjective views. Examine topics to include the genocide inflicted on Native Americans, the incarceration of Japanese Americans, and the enslavement of Black Americans. Seek trustworthy sources, synthesize the information, and decide for yourself how history shapes present-day policies and practices and how you'll respond to improve equity and inclusion. Knowing the truth will help us recognize false narratives and hidden agendas. Spend time on critical thinking, as discussed in Activity 17, "Devote Time to Critical Thinking."

Revisit Your *Why*

Staying connected to your *why*, as discussed in Activity 2, "Connect with Your *Why*, Find Your *Why Now*," is critical to successfully progressing on your journey to advance equity and inclusion. The never-ending road will be challenging at times, and it matters how we make people feel along the way. As we acquire knowledge of our nation's origin and connect the dots on how much organizations are built on systems of inequity, there's bound to be strong differences of opinion and emotional responses in interactions. Be prepared. Commit to open discussion where everyone can be heard and understood. Review Activity 28, "Reframe Difficult Conversations on Polarizing Topics." Endeavor for respect of others as you affect change.

Commit to the Process

Understanding history is essential to a better future. Acquiring knowledge about our historical origins and applying its lessons will require a process of continuous learning and emotional resilience. As we discover our nation's real history and how we came to be, long-held beliefs will be challenged, and a plethora of emotions will emerge that may include the rejection of the facts. I encourage you to keep an open mind and forge ahead. If you find yourself leaning toward rejection of the facts, that's the cue to explore the *why* around the rejection and to be honest with yourself. Invite someone with a different point of view to help with the exploration. Leverage diversity of thought. Emotions can serve as supporters or derailers of our journey, so we must reflect on all emotional reactions and their basis and distinguish between a past that cannot change and the interpretation of the past, which is always changing. Develop a list of resources and share with others on the journey. Consider the work of thought leaders including Nikole Hannah-Jones, Eduardo Galeano, Roxanne Dunbar-Ortiz, and Catherine Choy.

Action Accelerators

- **C-span.org**: "After Words," Nikole Hannah Jones: www.c-span.org/video/?515705-1/after-words-nikole-hannah-jones
- **NPR.org**: "Making the Case that Discrimination is Bad for Your Health": www.npr.org/sections/codeswitch/2018/01/14/577664626/making-the-case-that-discrimination-is-bad-for-your-health
- **YouTube.com**: *Roxanne Dunbar-Ortiz with Nikkita Oliver An Indigenous Peoples' History of the United States*: www.youtube.com/watch?v=m48UN4t-iLY
- **YouTube.com**: *Author Catherine Choy on the Lessons of Asian American History*: www.youtube.com/watch?v=ZtuqCc3ph10
- **NYT.com**: "In a Race to Shape the Future, History Is Under New Pressure," by Max Fisher (subscription required): www.nytimes.com/2022/01/05/world/history-revisionism-nationalism.html

Sources Cited

Janice Gassam Asare. 3 Ways Intergenerational Trauma Still Impacts The Black Community Today, Forbes.com, February 14, 2022, www.forbes.com/sites/janicegassam/2022/02/14/3-ways-intergenerational-trauma-still-impacts-the-black-community-today/?sh=640c9a6e3cf6

PubMed.gov. The weathering hypothesis and the health of African-American women and infants: evidence and speculations, *Ethn Dis*, 1992 Summer;2(3):207–21, pubmed.ncbi.nlm.nih.gov/1467758

Ana Sandoiu. "Weathering": What are the health effects of stress and discrimination? MedicalNewsToday.com, February 26, 2021, www.medicalnewstoday.com/articles/weathering-what-are-the-health-effects-of-stress-and-discrimination#An-apt-and-troubling-metaphor

Arline T. Geronimus, Margaret Hicken, Danya Keene, and John Bound. "Weathering" and Age Patterns of Allostatic Load Scores Among Blacks and Whites in the United States, *American Journal of Public Health*, ajph.aphapublications.org/doi/pdf/10.2105/AJPH.2004.060749

EEOC.gov. Jackson National Life Insurance to Pay $20.5 Million to Settle EEOC Lawsuit, www.eeoc.gov/newsroom/jackson-national-life-insurance-pay-205-million-settle-eeoc-lawsuit

EEOC.gov. JL Schwieters Construction to Pay $125,000 To Settle EEOC Race Harassment Lawsuit, www.eeoc.gov/newsroom/jl-schwieters-construction-pay-125000-settle-eeoc-race-harassment-lawsuit

Lynn Pasquerella. The Whitewashing of American History, AACU.org: Summer 2021, www.aacu.org/liberaleducation/articles/the-whitewashing-of-history

Be Willing to Learn from and Be Influenced by Others Who Are Different from You

"You are here to make a difference, to either improve the world or worsen it. And whether or not you consciously choose to, you will accomplish one or the other."

—**Richelle E. Goodrich**

***

Our environments have a tremendous influence on us and can impact just about every decision we make. Aside from DNA and upbringing, our environment informs the way we think, who we associate with, our likes and dislikes, and, more important, how we interact with one another. Consider a work culture that embraces the LGBTQ community. Such an environment signals that they can be their authentic selves, can fully engage, and will be supported in the face of harassment or discrimination. An environment that is intolerant of LGBTQ individuals begats intolerance and is an open invitation for others to display bias and homophobic tendencies signaling to conform to heterogeneous norms or else. In both scenarios, actions are influenced by the environment, bias, and stereotypical beliefs. But let's just say that in that intolerant situation, there are one to two individuals who model inclusive behaviors toward LGBTQ co-workers. They can influence change for those with an open mind to learn and make a conscious choice to be influenced. *Harvard Business Review* describes influence as:

> a neutral term indicating a person's capacity to have an effect on another person. Simply put, influence is what can move or sway someone to a desired action. Influence on its own is neither negative nor positive. You can influence others for good causes, such as influencing your family and friends to adopt fairness, justice, and honesty by displaying those behaviors yourself.

The people whom we are most influenced by tend to be those who are very similar to the way we are. That's our comfort zone. Whenever we seek guidance, search for fresh ideas, or want to gripe about something, we tend to seek those who think like we do, come from the same background, and share similar work-life experiences. They fuel or reinforce our biased positions and preconceived notions. Nothing feels better than feeling heard, understood, and validated by the people we rely on for acceptance and socialization. Our sphere of influencers conditions us to expect alignment with our perceptions and worldview such that when a disconnect occurs, we become defensive or discount their opposing point of view. When conversations venture down an unexpected path, the default is to argue that our perspective is the right perspective. While we know that a different point of view isn't a bad thing, it's not necessarily what we need or want to hear at the time. It goes against the grain and perhaps reveals an aspect of ourselves that we're not ready to face. However, ignoring them is a missed opportunity for learning and growth.

One sunny Saturday afternoon, while having lunch with close friends, I started a bitch session. Work life had been insane for weeks, and I needed to vent. I was feeling attacked as the only Black woman on the team. It felt really good letting it all out, unencumbered by political correctness and in the comfort of knowing that I was letting it all hang out in a circle of trust. My friends range in age, social status, and race, and we are very supportive of one another. They listened intently as we ordered another round of martinis. They chimed in from time to time with emotional reactions of, "Oh no!" "Really?" and "Girl pleeezzzz." As the conversation continued, the better I felt. One friend was just as shocked and appalled by the circumstances as I was. Another encouraged me with her philosophy of "trouble don't last always" and advised that I meditate to manage the stress and start looking for another job if it continued. But one friend, turned the tables when she asked me more questions and helped me realize my accountability in the situation. I actually became a little offended. How was any of this my fault? As much as I hate to admit it, her perspective was valid. She helped me see that there were things that I could have done differently, that I had made unfounded assumptions, and that I needed to consider the perspectives of the colleagues involved during those challenging weeks at work. It didn't mean that race wasn't a contributing factor as that's always a possibility, but I needed to be more strategic in my approach with the co-workers. I grew from her advice and continue to apply those lessons to this day. While the echo chamber was comforting, what occurred outside it was better for me in the long term.

The company we keep can have a positive or negative influence on how we build inclusive environments. The great thing is that we get to choose our influencers. As we strive to be more inclusive, we must continuously evaluate who and what informs our thinking, how well they are serving our goals, and paying attention to which voices are missing. Influencers can come from anywhere. In addition to friends, family, and co-workers, they may be historical or religious figures, or other individuals we've never met but either inspire us to be better versions of ourselves or, worse, reinforce our fears and/or racist, sexist, and stereotypical views. Our inclusion journey requires stepping outside the echo chamber and to not only seek diverse perspectives but to allow them in for further unbiased exploration and the possibility of adoption should they support our ambitions. With this in mind, reflecting on those who have affected the person we are today, we can make more conscious decisions about which influences we want to continue having in our lives. We've seen how a single individual can influence the masses. Consider Mahatma Gandhi and Martin Luther King Jr., who inspired civil rights and freedom movements, or Ibram X. Kendi, who is leading the charge on antiracism. By the same token, we've seen others like Adolf Hitler and Donald J. Trump who influence the masses with messages of bigotry and hate, which defend or justify systems of inequity. We can't negate the power of influence and its effect on shaping our societies, global political systems, and the workplace. Allowing ourselves to learn from and be influenced by others helps us to either exclude or include people of difference. Champions of inclusion seek influencers who help us become better allies, mentors, coaches, sponsors, and supporters of historically marginalized groups overall; in doing so, we give back by influencing others. We have the power to change old belief systems and be changed in the process.

Actions

Examine Equity and Inclusion from Multiple Perspectives

No single individual has all the answers. With their personal and professional experiences, influencers have unique approaches and philosophies on stamping out systems of inequity and calling out the issues. It is our job to understand those systems and how we perpetuate them. We benefit from their varied perspectives to aid us in the processing and interpretation of the barriers and challenges so that we become more effective at not only fighting *against* exclusion but *for* inclusion and equity. Here's a list of a few thought leaders who have influenced me with their diversity, equity, and inclusion (DEI) body of work. I encourage you to do your research. There are thousands to choose from throughout history and present day. Consider two to three individuals to follow on social media, or seek them out on podcasts, webinars, TED talks, events, booksellers' sites, and museum exhibits. Every few months add another to your list for continued growth and enhanced perspective.

- **Stacey Abrams:** American politician, lawyer, voting rights activist, and author
- **Vernā A. Myers:** Diversity and inclusion consultant, author
- **Ruchika Tulshyan:** Inclusion author and speaker
- **Ibram X. Kendi:** Professor, author, antiracist activist
- **Ta-Nehisi Coates:** Journalist, author
- **Alexandria Ocasio-Cortez:** Politician and activist
- **Ruth Bader Ginsburg:** United States Supreme Court Justice, gender equality advocate
- **Kimberlé Crenshaw:** Civil rights advocate, critical race theory scholar
- **James Baldwin:** Writer, civil rights activist
- **Jennifer Brown:** Diversity and inclusion consultant, author
- **Megan Rapinoe:** Soccer player and activist

Understand Your Power as an Influencer

As we look to others from whom we can learn on our inclusion journey, others are looking at us to discern our value as their influencers. We must be mindful of the signals we are sending. Checking in with ourselves to ensure that we are behaving inclusively even when no one is looking speaks volumes about our commitment. Use this awareness to take a more active role in influencing inclusive cultures.

Evaluate Your Influencers

List people who have had a significant impact on you and your inclusion journey. Think about whether they support or share your point of view and how they lend themselves to your ability to be more inclusive. Note how they affect your actions, feelings, and thinking. Consider whether there are voices and perspectives that you are not yet in tune with, i.e., people with disabilities, lesbians, or baby boomers. The wider our circle of influencers, the better we understand the issues to be a better ally, accomplice, or co-conspirator of change. By looking at specific stories of different individuals and situations, we can challenge our assumptions and test our DEI values.

Action Accelerators

- **PsychologyToday.com:** "How to Use Power, Influence, and Persuasion for Good," reviewed by Craig B. Barkacs: www.psychologytoday.com/us/blog/power-and-influence/202101/how-use-power-influence-and-persuasion-good

- **PsychologyToday.com:** "The Importance Of Good Influences," by Alex Lickerman: www.psychologytoday.com/us/blog/happiness-in-world/201004/the-importance-good-influences#:~:text=The%20more%20good%20influences%20with%20which%20we%20surround,created%2C%20which%20will%20also%20add%20to%20our%20happiness

- **Forbes.com:** "9 Core Behaviors Of People Who Positively Impact The World," by Kathy Caprino: www.forbes.com/sites/kathycaprino/2014/06/02/9-core-behaviors-of-people-who-positively-impact-the-world/?sh=407c4bb46b41

Sources Cited

Ruchi Sinha. Are You Being Influenced or Manipulated? HBR.org, January26,2022,hbr.org/2022/01/are-you-being-influenced-or-manipulated

52

Honor Your Commitment to Inclusion

"We have no hope of solving our problems without harnessing the diversity, the energy, and the creativity of all our people."

—**Roger Wilkins**

***

Inclusion is a verb. I love the phrase coined by the Winters Group, "Diversity is the mix; inclusion is making the mix work," as well as Vernā Myers' quote, "Diversity is being asked to the party, inclusion is being asked to dance," and my maxim, "Inclusion is everybody's job that anybody can do." All three require that we must "do something" before inclusion can begin. Cultures that embrace difference are influenced not only from the top of the organization down but from the bottom up. The passion, power, and purpose within us should never be aimed at perpetuating systems of oppression, racial, or social injustice. We must ask ourselves why we would want to do such a thing. We know it's not fair. We know it causes harm. We know that we would not want this for a loved one or ourselves personally, and we know that people are dying because of it. We spend the majority of our lives in the workplace. For at least 40 hours per week, we can either make a difference/be the difference for a co-worker or build a barrier/be the barrier. Each day we get to choose.

As we strive for equity and inclusion for all, we must keep our inclusive lenses pure—that is, free of bias, full of authenticity, and always close by and ready at a moment's notice. Exclusion never takes a break, and neither can we. Inclusive behavior cannot be something that we turn off at the end of the work or school day. While it takes emotional energy to ignore, justify, tolerate, or even care that a racist joke, for instance, was just shared by a co-worker or family member, it takes action to speak up against it and prevent or minimize it from happening in the future. Without the vision of workplaces and societies where historically excluded groups have an equal opportunity to thrive rather than scrape by or worse perish, roadblocks will seem insurmountable, and we'll return to old habits. Staying connected to the work means evolving with societal shifts, maintaining an open mindset, and recognizing that what worked last year may not be as effective a year or two from now. Never think that because the trauma, tragedies, and prejudicial acts of co-workers experienced by marginalized groups aren't happening to you or don't affect you that they won't in the future. Covid affected everybody. The reversal of *Roe v. Wade* affects all women to varying degrees regardless of race and socioeconomic background. A parent may be against LGBTQ rights until faced with knowledge that their child is gay. Members of the dominant culture need never be concerned about a loved one surviving a police traffic stop or other racial injustices until they have a biracial family member. Perhaps you've noticed a colleague being unfairly singled out by the boss consistently. You're making a grave mistake thinking that silence buys security. Trust and believe that there is nothing preventing the boss's behavior from turning toward you. Imagine the impact that could have been made had you started this journey years ago. The fight for equity has been going on for centuries. When we are determined to do or have something, we let nothing get in our way. The energy just shows up. People put their resources where their passions lie. I've seen people camp outdoors overnight in freezing temperatures to be the first in line for a sports event ticket. Consider the astonishing act of Patagonia founder Yvon Chouinard, who gave away his $3 billion outdoor apparel retail empire to a trust and nonprofit to help combat climate crises. Imagine leveraging whatever level of privilege you have to influence change on some level. Then consider how much further we could be or what reprehensible acts could have been avoided. Imagine the possibilities of individual and collective acts over time. If we are sincerely striving to be a part of the solution, there are no more excuses. Everybody is capable of doing something. I know that it takes courage to overcome the fear of getting it wrong. Consider the tragedy of doing nothing in favor of convenience. Face that fear, and do something anyway. It will take resilience to bounce back after a setback as well as commitment to never give up. Whenever we commit to anything worthwhile, we have a much better chance of success when we persevere, learn from mistakes, and lean on one another when needed. Reflect on your greatest achievements; more than likely the road was not linear, and there may have been numerous temptations to give up. Staying the course allows us to continuously grow as individuals, become more empathetic, and turn that empathy into action as we create and nurture relationships. The work we do today lays the foundation for the next generation to carry forth. This is a lifelong journey that we practice daily and teach our children. They are the workplace of the future. If there is no one in your sphere of influence leading the inclusion charge or modeling inclusive behaviors, that means that there's a void that only you can fill. Remember, doing nothing is doing something—condoning the status quo. Champions of inclusion are beyond talking about the issues. We are doing something about it—daily.

Actions

Lead the Way

It's within everyone's capacity to create environments of inclusion and belonging every day. You don't have to be in a leadership role to lead the way. You may be the only one on your team or in your company who uses appropriate pronouns, says and spells names correctly, or speaks up in the face of prejudice, sexism, or racism. Apply the 52 activities regularly, and set the example for others to follow.

Endeavor to Improve Every Day

Nothing ventured, nothing gained. No pain, no gain. The definition of insanity is doing the same thing over and over expecting a different result. I'm not sure of the origination of these maxims, but one thing is for sure: each one of them make sense. Continuous improvement is the only way to build inclusion muscle and bring about lasting change. Assess how you're doing every fourth activity; that's about once a month. Are there lessons learned that can be carried into the next four activities? Reflect on anything that did not go as you hoped, and evaluate what could be done differently in the future. Perhaps you have become braver and can now act, where before you were afraid. Consider whether you owe someone an apology. Small things add up. It's never too late to do the right thing.

Plan to Succeed Together

Teamwork makes the dream work. Everybody needs somebody sometime. Each one, teach one. OK, no more catch phrases. I promise. Working together can only accelerate results. Collective wisdom focused on the shared goal of creating more equitable and inclusive workplace cultures and societies is a winning hand. Be intentional about working with others, and remain vigilant in bringing more people in. We're not at the same place on this journey, so we must meet people where they are. Build a groundswell that can't be ignored. The journey is the destination.

Begin Again

Congratulations on completing all 52 activities. Don't put the book on the shelf now. Start again at Activity 1, "Know Thyself." If you've been journaling your journey, review it and see how much you've progressed. Exclusion didn't stop when you reached the final activity. Year 2 gets You.v2. I'm excited to hear all about it. Live by your inclusive values. Join us on LinkedIn, and share your successes in the group "Champions of Inclusion."

Action Accelerators

- `LegalManagement.org`: "12 Ways to Commit to Diversity and Inclusion," by Michelle Silverthorne: `www.legalmanagement.org/2020/january/columns/12-ways-to-commit-to-diversity-and-inclusion`

- `Forbes.com`: "Commit to Inclusion - Establish Anti-Racist Team Norms," by Sabina Nawaz: `www.forbes.com/sites/sabinanawaz/2021/06/08/commit-to-inclusion-establish-anti-racist-team-norms/?sh=630bc50b2c12`

- **Podcasts:** *Inclusion Begins with Me: Conversations That Matter* (view all eight episodes), MetLife: podcasts.apple.com/us/podcast/inclusion-begins-with-me-conversations-that-matter/id1613461389

- **WomenTakingtheLead.com:** "Jennifer Brown on the Personal Journey to Be Inclusive": womentakingthelead.com/jennifer-brown

Index